Advances in Anatomy Embryology and Cell Biology

Vol. 160

Springer
Berlin
Heidelberg
New York
Barcelona
Hong Kong
London
Milan
Paris
Singapore
Tokyo

R. Sharma U. Schumacher

Carbohydrate Expression in the Intestinal Mucosa

With 32 Figures and 8 Tables

Springer

Ram Sharma
University of Southampton
Human Morphology School of Medicine
Biomedical Sciences Building
Bassett Crescent East
Southampton SO16 7PX, UK
E-mail: rps2@soton.ac.uk

Udo Schumacher
Universitätsklinikum Hamburg-Eppendorf,
Institut für Anatomie
Abteilung für Neuroanatomie,
Martinistraße 52,
20246 Hamburg, Germany
E-mail: uschumac@uke.uni-hamburg.de

ISSN 0301-5556
ISBN 3-540-41669-2 Springer-Verlag Berlin Heidelberg NewYork

Library of Congress-Cataloging-in-Publication-Data
Die Deutsche Bibliothek – CIP-Einheitsaufnahme
Sharma, Ram: Carbohydrate expression in the intestinal mucosa / R. Sharma and U. Schumacher. – Berlin; Heidelberg; New York: Springer, 2001
(Advances in anatomy, embryology, and cell biology; Vol. 160)
ISBN 3-540-41669-2

Springer-Verlag a member of BertelsmannSpringer
Science + Business Media GmbH

http://www.springer.de

Production: PRO EDIT GmbH, 69126 Heidelberg, Germany
Printed on acid-free paper – SPIN: 10791530 27/3136wg - 5 4 3 2 1 0

Contents

Abbreviations

AB	Alcian blue
AB-PAS	Alcian blue — periodic acid Schiff
ABC	avidin biotin complex
APUD	amine precursor uptake and decarboxylation
CCK	cholecystokinin
Cv	conventional
CV-P	conventional animal fed purified diet
CV-C	conventional animal fed commercial diet
DAB	diaminobenzidine
EC	enteroglucagon
FAE	follicle-associated epithelium
FITC	fluorescein-isothiocyanate
GALT	gut-associated lymphoid tissue
GF	germ-free
GF-P	germ-free animal fed purified diet
GF-C	germ-free animal fed commercial diet
GLI	grey level index
H&E	haematoxylin and eosin
HFA	human flora-associated
HFA-P	human flora-associated fed purified diet
IBAS	interaktives Bild Analyse System
PAS	periodic acid Schiff
PYY	peptide with tyrosine (Y) at the N-terminus and tyrosine (Y) amide at the C terminus
SCID	severe combined immunodeficiency
SCFA	short chain fatty acid
SIR	serotonin immunoreactive
-	Abbreviations for lectins and their specific sugars are shown in Tables 2, 3 and 4

1 Introduction

The mucosal surface of the gastrointestinal tract is covered by a monolayer of columnar epithelial cells. This epithelium represents a vast surface area that is vulnerable to foreign antigens and microbial pathogens. By being in contact with a large number of potentially harmful substances and infectious organisms, the mucosal surface must provide mechanisms not only to regulate the entry of macromolecules but to serve as an exporter of secretory antibodies for mucosal defence. Consistent with these functions, the epithelial surface of the gastrointestinal tract is lubricated and protected by mucus secretion and by a highly specialised immune system underlying the epithelium which exports the appropriate immunoglobulins into the intestinal secretions.

The experimental work presented in this monograph has targeted the mucin-type glycoproteins and gut-associated lymphoid tissues as selected areas of research to explore their roles in barrier function and immune defence of the intestinal mucosa.

1.1 Intestinal Mucins

The term mucus, although loosely applied to the slimy fluid of the intestinal lumen, occurs partly as a stable water-insoluble semisolid gel adherent to the epithelial surface, and partly in a water-soluble form in the intestinal lumen (Forstner 1978). The viscous and elastic properties of the mucus gel are attributed to the polyanionic structure of mucus glycoproteins and their physical interactions with other molecules (Specian and Oliver 1991). These mucus glycoproteins, now universally referred to as the mucins or mucin-type glycoproteins, are synthesised and secreted by the goblet cells dispersed among the enterocytes within the epithelium of the gut and its glandular derivatives (Neutra and Forstner 1987). Mucins secreted from the goblet cells and exfoliated enterocytes subsequently mix with the bacteria, mucopolysaccharides and water on the epithelial surface of the intestinal mucosa (Mantle and Allen 1989).

The epithelial structures of the mucosa and the chemical characteristics of mucins along the gastrointestinal tract reflect the regional differences in the absorptive and secretory activity of its various segments. Since the functional modulation of gastrointestinal mechanisms is closely associated with the structural alterations of the mucosa and its mucins, the changes in mucosal architecture and histochemical composition of mucins have received increasing attention since these are linked to the pathogenesis of various inflammatory and neoplastic gastrointestinal disorders (Forstner et al. 1982; Jacobs and Huber 1985).

This epithelial surface is subjected to dynamic alterations by intraluminal dietary constituents and numerous species of bacteria (Heneghan 1979; Vahouny et al. 1985). In order to maintain the protective functions of intestinal mucosa and thickness of the mucosal barrier, the loss of mucus and enterocytes from the luminal surface must be balanced by the synthesis of mucins and renewal of both goblet cells and enterocytes from the crypts (Allen 1981). The rapid turnover of epithelial cells and the continuous renewal of mucins make the intestinal mucosa an ideal model system for studying the effects of interactions between gut organisms and food constituents on cell proliferation and mucus secretory mechanisms.

The task facing the investigators in the field of gastroenterology is to analyse the nutritional and bacterial factors that affect the integrity of mucosa, alter the production of mucin and its chemical composition, or enhance its degradation. The understanding of protective mechanisms will help to develop rational approaches towards effective treatment of intestinal disorders.

1.2 The Mucosal Immune System

The intestinal lumen is replete with foreign matters and upon any encounter with antigenic stimuli, the epithelial cells must deliver samples of luminal antigens to the mucosal immune system without flooding the systemic circulation with antigens. The gut-associated lymphoid tissue (GALT) represents a highly compartmentalised immunological system and serves as a principal mucosal inductive site where immune responses are initiated (Kraehenbuhl and Neutra 1992). Antigens, particles and microorganisms are transported across the epithelial barrier by membranous cells (M cells), which unlike most epithelial cells, are structurally and biochemically specialised for sampling and vesicular transport of endocytosed materials (Gebert 1997). After M-cell transport, antigens are channelled to the microenvironment of the GALT, where they are processed by macrophages, dendritic cells and B cells, resulting in an immune response. This response is mediated by B cells which proliferate locally in the GALT and migrate systemically to distant mucosal and glandular tissue, where they differentiate into IgA-producing plasma cells (Giannasca and Neutra 1994).

The sampling of antigens by M cells as the prerequisite for mounting a secretory immune response has inherent risks in that pathogens can exploit the M-cell transport mechanisms to invade the host. However, the binding of antigens to the M-cell surface may be a key step in achieving mucosal immunity and therefore an understanding of the receptors on M-cell apical membrane should be vital in the design of oral vaccines (Neutra et al. 1996).

1.3 Objectives and Scope of the Research Presented in This Monograph

In view of the relevance of mucins in the pathogenesis of several gastrointestinal diseases there is a compelling need to obtain a clear insight into the mechanisms responsible for their biosynthesis, storage, secretion and degradation. Much of our knowledge of mucin-like glycoproteins has been derived from biochemical and molecular studies, which reflect the properties and behaviour of intestinal mucins, but give meagre information regarding the factors that alter the synthesis and secretion of mucins. To consolidate the understanding of the physiology of mucus, the work presented in this monograph has focused on the mucosal responses to dietary constituents in the intestinal lumen and on the microorganisms that colonize the epithelial surface. The experimental work was designed to resolve whether the diet and/or microbial flora affect the intestinal morphology and/or the nature and the distribution of the carbohydrate residues in the mucus-secreting cells. The experimental model system was further utilised to evaluate the effects of luminal challenge of diet and/or microbial flora on the distribution of enteroendocrine cells along the intestinal tract.

Since M-cell transport mechanisms are exploited by pathogenic organisms to gain access to target cells in the intestinal mucosa, an understanding of specific receptors on M cells is vital for the design of oral vaccination strategies. Mucosal immunologists frequently ask the following questions: How do bacteria recognise and attach to M cells? Are there any apical membrane components on M cells that might serve as potential receptors? Why is the M-cell surface more accessible to luminal pathogenic organisms than the neighbouring enterocytes? To answer these questions recent studies have focused on the mechanisms that may allow M cells to sample pathogens, either by the binding of lectin-like bacterial adhesions to M-cell surface glycoconjugates or conversely, the binding of bacterial surface carbohydrates by M-cell surface lectins. The studies presented in this monograph concentrate on the glycoconjugate characterisation of the mouse and human GALT and on the overlying follicle-associated epithelium (FAE), in order to identify their cell-specific carbohydrate residues.

2 Materials and Methods

2.1 Animals

Twenty-five 9-week-old male rats were used for the study. The germ-free animals (GF; n=15) were born and maintained in plastic isolators while the conventional controls (CV; *n*=10) were kept in the open laboratory. Five of the germ-free rats were inoculated 1 week after weaning (i.e. at 4 weeks of age) with 1 ml of 20% suspension of human faecal organisms equivalent to the administration of a total bacteria count of 5×1010 microorganisms by oral intubation (HFA; see Fig. 1). At the end of the experiment the caecal contents of HFA animals were subjected to an enzyme analysis for β-glucosidase, β-glucuronidase, nitrate reductase and nitroreductase to confirm the human flora were still present (Mallett et al. 1987).

2.2 Diets

Two nutritionally adequate diets, which differed markedly in constituents and in texture, were used (Table 1). Five germ-free (GF-C) and five conventional (CV-C)

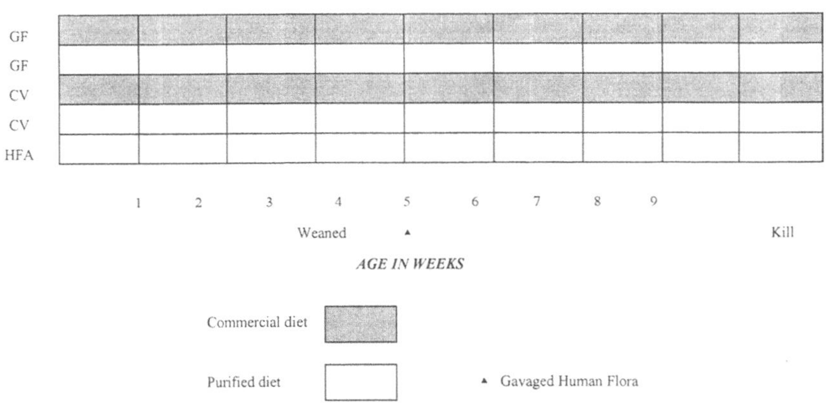

Fig. 1. Microbial status and diet assignment of animal groups

Table 1. Composition of diets

Purified diet (P)	
Component	Concentration (g/kg)
Casein	250
Maize starch	380
Potato starch	100
Sucrose	50
Cellulose powder	80
Soya bean oil	60
Mineral mix	60
Vitamin mix	20
Commercial Diet (C)[a]	
Cereal products (barley, maize, wheat and wheatfeed)	775
Extracted soya bean meal	107
Fish meal	98
Mineral and vitamin supplement	20

[a]Calculated crude fibre content 4.1%.

animals were fed on a commercial rat diet (diet GR3; Special Diets Service, Witham, Essex); the fibre content of this diet was derived from coarsely milled cereal products. Five germ-free rats (GF-P) and five conventional rats (CV-P) were maintained in the laboratory on a diet prepared in the laboratory from purified ingredients that contained 8% cellulose as the source of fibre (Table 1). Both diets were sterilised by γ-radiation at 50 kGy. The five HFA rats also received this diet.

2.3 Preparation of Tissues and Histochemical Methods

Since a 4-week diet and microflora treatment period in post-weaned rats is adequate to detect changes in the mucosa (Meslin et al. 1993), animals were killed at 9 weeks of age. Three-centimetre lengths of small intestine (jejunum) were taken at 15 cm distal to the gastropyloric junction and of the proximal large intestine at 2 cm from caecum. All samples were fixed in 10% buffered formalin and embedded in paraffin wax.

Tissue blocks were serially sectioned at a thickness of 5 μm using a Reichert rotary microtome. To minimise errors due to section thickness, sections were stained with haematoxylin and eosin (H&E) and any sections with overlapping nuclei in villus epithelium were discarded.

Sections were subjected to the following procedures for the identification of mucosubstances:

1. The periodic-acid Schiff (PAS) reaction for studying unsubstituted α-glycol-rich neutral mucins.
2. 1% Alcian blue, pH 2.5, followed by PAS (AB 2.5/PAS) to allow neutral (pink) and acidic (blue) mucins to be differentiated. With this procedure a purple colour is obtained when neutral mucins are also present, and deep purple when neutral mucins are mainly present (Filipe 1979).
3. 1% Alcian blue, pH 1.0 (AB 1.0), for the selective characterisation of sulphomucins (Lev and Spicer 1964).

The Alcian blue dyes and the techniques used in this study were pretested in a dot blot assay for their suitability to label carboxylated and sulphated mucins, respectively (Schumacher and Adam 1994).

Semiquantitative staining intensities were based on a scale ranging from 0, unreactive, to +++, intensely stained.

2.4 Quantitative Morphology

Ten longitudinally oriented villi and crypts from the small intestine and ten crypts from the large intestine were selected randomly from each animal for measurement. Villus length was measured from the tip to the base of villus and the crypt length was measured from the bottom of the crypt to the opening of the crypt (Sigleo et al. 1984). The results were expressed as mean ± standard error of the mean (SE) and statistical differences between groups were analysed for significance by Student's *t*-test.

2.5 Morphometry

Computerized morphometric measurements were carried out using a Zeiss photomicroscope with an eyepiece of magnification 10 and an objective of magnification 16. The image was processed for analysis on a TV-based IBAS image analysis system (Kontron, Watford, Herts, UK). The system is centred around the central processor unit which was used for grey level index (GLI) measurement to include the objects on the basis of their colour and intensity, and for enhancing and storing the GLI images. A video camera mounted on the microscope was used as an input device and the interactive manipulation was done on the colour images of microscopical fields displayed on the monitor by using a digitising tablet.

2.5.1 Measurement Protocol

From each animal histological sections stained with PAS, AB pH 2.5 and AB pH 1.0 were used. To ensure comparability between the different groups of animals, the

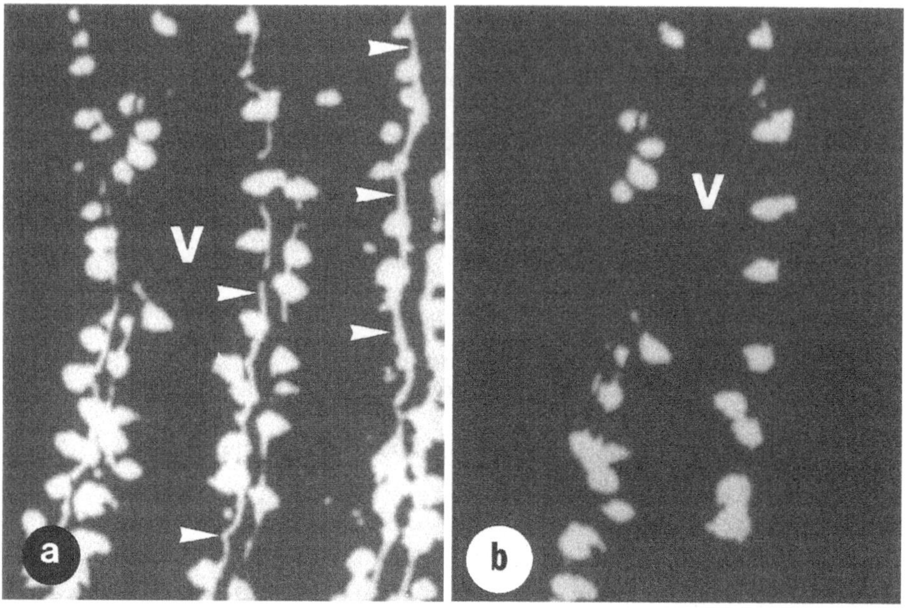

Fig. 2a,b. Image analysis of small intestinal villi in a germ-free rat fed a commercial diet. **a** The binary image of the positive areas representing intestinal mucins are in *white* and negative areas in *black*. The surface mucins are *arrowed*. **b** Results of editing the surface mucins and extraction of positively stained mucin granules in a villus. Measurements were made on the intestinal mucins of one extracted villus (*V*)

sections from all experimental groups were stained in one batch so that differences in the staining pattern can not be attributed to technical manipulations. Two blocks of tissue from each animal were sectioned and ten longitudinally oriented villi from the small intestine and ten crypts from the large intestine were selected randomly from each animal for analyses; 50 villi and crypts per experimental group were analysed.

The villi and crypts were outlined on the monitor screen by using the digitising tablet and the area was recorded in square micrometers (μm^2). The profiles of histochemically stained mucins of a villus or a crypt were enhanced by adjusting the GLI of the image analyzer. As the histochemical procedures employed in this study stain the mucus granules of goblet and crypt secretory cells, the apical secretion of these cells as well as the surface mucus are all included in the image since the range of their grey level corresponds with the GLI of goblet cells. In order to ensure that only measurements of mucin-containing cells were computed on the image analyzer, editing was used to eliminate the areas which were staining the apical secretion. Figure 2 shows the process of colour image analysis. In the binary image the positive areas representing mucus granules were displayed in colour and the negative areas in black. The number of pixels in the positive areas were computed automatically to determine three field measurements of the villi or crypts: (1) the total number of mucin-containing cells; (2) the total volume density of mucin granules; (3) the density of stained mucin granules.

The number of mucin-containing cells/mm^2 of villus/crypt cross-sectional area was calculated by multiplying the mean number of mucin-containing cells/villus or crypt with 10^6 divided by the mean crypt/villus cross-sectional area. The staining densities of intracellular neutral (PAS), carboxylated (Alcian Blue, pH 2.5) or sulphated (Alcian Blue, pH 1.0) mucin granules in the goblet and crypt secretory cells generated by the scanning procedure were measured in GLI units by the IBAS system. All data in this study were generated by a single investigator who also calibrated the system which included the appropriate microscope set-up, adjustment of the light source, acquiring the blank field and performing the shading correction of the image. In order to avoid bias due to the method, the section thickness (5 μm) and the staining schedule of histochemical reactions were standardised. The sections were numbered in code, blindly scanned from the experimental groups to eliminate further bias, and the code broken at the end of experiment.

2.5.2 Statistical Analysis

The results are expressed as mean ± standard error of the mean. Student's *t*-test for unpaired data was used for comparison of mucin-containing cell numbers between two groups. The differences between the volume densities of mucin granules and the staining densities of mucin granules between groups were evaluated by a two-sided normal *t*-test (Kirkwood 1992). Differences were considered significant at the $P<0.05$ level.

2.6 Lectin Histochemistry to Identify Carbohydrate Residues of Mucin

For identification of specific carbohydrate residues, the wax sections were dewaxed in xylene and rehydrated through a series of alcohols and brought to 0.05 mol/l TBS (Tris-buffered saline, pH 7.4) supplemented with 0.1% calcium chloride. Sections were then incubated for 30 min at room temperature in a Shandon Sequenza immunostaining centre with various FITC-conjugated lectins (Sigma, Poole, Dorset, UK, or E.Y. Laboratories through Bradsure Biologicals, Market Harbourough, Leic, UK) using a 1:20 dilution of a 1 mg/ml solution in TBS. For lectins, their abbreviation and sugar specificity see Table 2.

2.6.1 Controls

In control experiments, the lectin binding was blocked by incubating (30 min, room temperature) the FITC-conjugated lectins with their appropriate inhibitory sugars (0.3 mol/l end concentration) before application to the sections. In addition, the sections incubated with buffer solution alone also served as controls for possible autofluorescence. After rinsing with the TBS, the sections were mounted in AF1

Table 2. Characteristics of FITC-conjugated lectins [adapted from Danguay et al. (1994) and commercial literature of Sigma Chemical and E.Y. Laboratories]

Lectin origin	Abbreviation	Carbohydrate binding specificity	Inhibitory sugar
Canavalia ensiformis	Con-A	α-d-Man > α-d-Glc	α-Mannose
Triticum vulgaris	WGA	GlcNAc(β1,4GlcNAc)$_{1-2}$ > β1,4GlcNAc > NeuAc	GlcNAc
Dolichos biflorus	DBA	GalNAcα1,3GalNAc > GalNAcα1,3Gal	d-GalNAc
Maackia amurensis	MAA	NeuAcα2,3Galβ1,4GlcNAc	[a]
Ulex europaeus	UEA-I	l-Fucoseα1,2Galβ1,4GlcNAcβ1,6	α-l-Fucose

Fuc, fucose; Gal, galactose; Glc, glucose, GlcNAc, *N*-acetylglucosamine; GalNAc, *N*-acetylgalactosamine; Man, mannose, NeuAc, neuraminic acid (sialic acid).
[a] Pre-treatment of the sections with neuraminidase was used.

mountant (Citifluor, City University, London, UK) and examined with an Olympus BH2 fluorescence microscope with the appropriate filters for FITC. Photographs were taken using Kodak TMax 400 ASA film.

2.6.2 Neuraminidase Digestion

Terminal sialic acid residues were removed by neuraminidase and the glycoconjugates with sub-terminal d-GalNAc were identified by staining with lectin DBA. To test whether enzyme-free buffer is as effective as neuraminidase in unmasking lectin receptors, the binding of DBA to goblet cells before and after neuraminidase treatment was investigated. Sections were pretreated with type X neuraminidase from *Clostridium perfringes* (Sigma) at 1 unit/7 ml according to the procedure by Plendl et al. (1989). The slides were incubated with neuraminidase at 37°C overnight, washed in TBS, and labelled with the lectin. Parallel experiments were performed using enzyme-free buffer. The results were compared with those of direct labelling of sections without any prior incubation.

Results were recorded according to the intensity of fluorescence in individual goblet cells and graded semiquantitatively from negative (0) in nonreactive cells to intensely fluorescent (+++). Photographs were taken using a Kodak Tmax 400 black and white film.

2.7
Staining Procedure for Neuroendocrine Cells

Serial 5-μm sections were stained with the Grimelius silver nitrate technique (Grimelius 1968) for the study of argyrophilia of endocrine cells. For immunocytochemical demonstration of serotonin the sections were stained by the avidin-biotin-complex (ABC) procedure (Hsu et al. 1981) using a primary antiserum raised in goat and conjugated to bovine serum albumin (Incstar, Wokingham, Berkshire, UK). Sections were dewaxed and placed in 100% ethanol, and endogenous peroxidase was blocked using 1.6% hydrogen peroxide in methanol. After rehydration, sections were washed in 0.15 M Tris-buffered saline, pH 7.3 (PBS), and incubated for 20 min in 10% nonimmune rabbit serum. Incubation with the primary antiserum (1:1000) in PBS/0.3% Triton X-100 containing 1% normal rabbit serum was carried out overnight at 4°C in a moist chamber. This was followed by a brief wash in PBS, and incubation for 30 min in the secondary antiserum (biotinylated rabbit anti-goat immunoglobulin, 1:200). After a further wash in PBS, sections were treated with ABC complex from peroxidase standard PK-4000 (Vector Laboratories, Peterborough, UK) for 30–90 min, washed again in PBS, and then incubated in 0.5% diaminobenzidine tetrahydrochloride (DAB) in Tris-HCl buffer, pH 7.3, with 0.001% hydrogen peroxide for 5–10 min.

2.7.1
Controls

(1) Use of nonmimmune goat serum in place of the primary antiserum; (2) pretreatment of antiserum with serotonin/BSA conjugate. Sections were lightly counterstained with Harris haematoxylin, mounted in DPX and examined by light microscopy.

2.7.2
Quantitative and Statistical Methods

The expression of argyrophilic reaction and serotonin was correlated with the position of cells occupied along the crypt-villus axis. Therefore the crypts and villi of jejunum and colon were divided into three domains: the lower half of the crypt, the upper half of the crypt and the surface epithelial cells. The number of Grimelius-positive and serotonin-immunoreactive endocrine cells per unit length of mucosa (4 mm), delineated by an eyepiece graticule, were counted with an eyepiece of magnification ×10 and an objective of magnification ×20 in sections of the jejunum and colon from each of the animals. Cells were randomly selected from 10–15 visual fields from each rat showing the longitudinally oriented villi and crypts along the entire length of the mucosa. A similar method to express the number of enteroendocrine cells per unit length of mucosa has been used previously (Sjölund et al. 1982). The results are expressed as means ± standard error of the mean. Student's *t*-test for unpaired data was used to evaluate differences between groups. Differences were considered significant at the $P<0.05$ level.

2.8
Syngeneic Bone Marrow Transplantation

2.8.1
Animals and Bone Marrow Transfer

Twenty-four 3- to 4-month-old pathogen-free BALB/c C57BL/Kalgh-I male scid/scid (SCID) mice were sampled from the breeding program of the University of Southampton Biomedical Sciences Building animal house. A laminar flow isolator was used during manipulations of the animals which were housed in groups of five animals in a filter top cage on a sterile bedding and provided with sterile food and water ad libitum. Ileal specimens (6–8 cm proximal to ileocaecal junction) from 12 4-month-old normal male BALB/c mice fed a standard rodent diet and maintained under conventional laboratory conditions from the same animal facilities were examined as controls.

For bone marrow transplantation, bone marrow was removed from the femora of adult male BALB/c mice. Each end of the femora was cut off and Hank's balanced salt solution (HBSS) was flushed through the bone using a 23G syringe needle and a 10-ml syringe. The harvested cells were centrifuged after washing in HBSS and counted. They were resuspended in HBSS and 0.1 ml containing 6.3×106 cells injected intravenously into the tail vein of each SCID mouse. Fifteen SCID mice were divided into groups of five and were killed 2, 4 and 6 weeks after transplantation and the remaining nine transplanted SCID mice were left for 10 weeks. The mice were killed by cervical dislocation and three to four pieces of the ileum containing lymphoid follicles were dissected from each animal using a binocular microscope. The tissues were rolled up like a Swiss roll, fixed in a solution of 4% paraformaldehyde in 0.1 M phosphate buffer (pH 7.4) for 18 h at 4°C and processed routinely for wax histology.

2.8.2
Sample Processing and Statistics

One-hundred serial transverse 5- to 6-µm sections were cut from Swiss-rolled ileal samples of each of the 12 Balb/c control mice and of the 24 reconstituted SCID mice. Every tenth equidistant section was stained with haematoxylin and eosin (H&E) and the number of single primary follicles and Peyer's patches were counted on ten H&E sections from each animal. Statistical comparisons between the groups were analysed using one-way analysis of variance (ANOVA; Tukey's multiple comparison test) and differences were considered significant at the $P<0.05$ level.

2.8.3
Lectin Histochemistry

Ten paraffin sections from each animal were deparaffinised in xylene, hydrated through a series of graded alcohols and brought to 0.05 M Tris-buffered saline (TBS) at pH 7.4 containing 0.1% calcium chloride. Sections were trypsinised with 0.1%

Table 3. Lectins used, their carbohydrate specificities and inhibitors [adapted from Danguay et al. (1994) and commercial literature of Sigma Chemical and E.Y. Laboratories]

Lectin	Abbreviation	Carbohydrate binding specificity	Inhibiting sugar
Canavalia ensiformis	Con A	α-d-Man > α-d-Glc	α-Mannose
Dolichos biflorus	DBA	GalNAcα1,3GalNAc > GalNAcα1,3Gal	d-GalNAc
Maackia amurensis	MAA	NeuAcα2,3Galβ1,4GlcNAc	[a]
Ulex europaeus	UEA-	α-l-Fucose-	Fucose
Anguilla	IAAA	α-l-2 Fucose	Fucose

Gal, galactose; Glc, glucose; GlcNAc, *N*-acetylglucosamine; GalNAc, *N*-acetylgalactosamine; NeuAc, neuraminic acid (sialic acid).
[a] Neuraminidase predigestion of the sections

trypsin (Sigma) in 0.05 M TBS for 30 min at 37°C and after a wash in TBS, the endogenous peroxidase activity was blocked with 0.3% hydrogen peroxide in methanol. Sections were washed again in TBS and incubated for 30 min at room temperature in a Shandon Sequenza immunostaining centre with the biotinylated lectins at a concentration of 10 μg/ml in TBS. Biotinylated α(1–2)-fucose-specific lectins from *Anguilla anguilla* (AAA; E.Y. Laboratories, San Mateo, Calif., USA) and *Ulex europaeus* type I (UEA-I; Sigma) were used to characterise M-cells. The names, sugar specificities and inhibitors of further lectins used in this study are listed in Table 3. After a further wash in TBS, sections were treated with ABC complex (peroxidase standard PK-4000; Vector Laboratories, Peterborough, UK) for 30 min, washed again in TBS, and incubated in 0.5% diaminobenzidine tetrahydrochloride (DAB) in Tris-HCl buffer, pH 7.3, with 0.3% hydrogen peroxide for 5–10 min. The control experiments were carried out by omission of the lectin and by incubation of the sections with the lectins and their appropriate inhibitory sugar at a concentration of 0.3 M except for MAA, in which a neuraminidase predigestion was used (Plendl et al. 1989). Sections were lightly counterstained with Harris haematoxylin, mounted in DPX and examined by light microscopy. Photographs were taken on Kodak Technical Pan black and white film.

2.9 Human and Mice Peyer's Patches

2.9.1 Intestinal Samples

Human ileal specimens were obtained as resection margins from Southampton General Hospital. Specimens containing Peyer's patches from the distal part of human small intestine judged by routine pathological and histological examination as normal were obtained from nine patients with carcinoma of the colon.

Ten adult (4- to 6-week-old) male BALB/c mice from the breeding program of the University of Southampton Biomedical Sciences Building animal house, weighing approximately 20–25 g, were used. They were fed a standard rodent diet and maintained under conventional laboratory conditions. Pieces of mid-intestine containing lymphoid follicles were dissected using a binocular microscope from mice anaesthetised with ether and killed by cervical dislocation.

All the tissues from human and mice intestine were fixed in neutral buffered formaldehyde and processed routinely for wax histology.

2.9.2 Lectin Histochemistry

Paraffin sections were cut at 5–6 µm, deparaffinised in xylene, hydrated through a series of graded alcohols and brought to 0.05 M Tris-buffered saline (TBS) at pH 7.4 containing 0.1% calcium chloride. Sections were trypsinised with 0.1% trypsin (Sigma) in 0.05 M TBS for 30 min at 25°C and after a wash in TBS, the endogenous peroxidase activity was blocked with 0.3% hydrogen peroxide in methanol. Sections were washed again in TBS and incubated for 30 min at room temperature in a Shandon Sequenza immunostaining centre with the biotinylated lectins (Sigma, E.Y. Laboratories, Vector Laboratories) at a concentration of 10 µg/ml in TBS. The lectins used in this study, their sources, abbreviations and sugar specificities are listed in Table 4. The recently isolated lectins were biotinylated using NHS-biotin long arm spacer (Vector Laboratories, Peterborough, UK) as described previously (Schumacher et al. 1993). After a further wash in TBS, sections were treated with ABC complex (peroxidase standard PK-4000; Vector Laboratories, Peterborough, UK) for 30 min, washed again in TBS, and then incubated in 0.5% diaminobenzidine tetrahydrochloride (DAB) in Tris-HCl buffer, pH 7.3, with 0.3% hydrogen peroxide for 5–10 min. The control experiments were carried out by omission of the lectin and by incubation of the sections with the lectins and their appropriate inhibitory sugars (0.3 mol/l end concentration). Sections were lightly counterstained with Harris haematoxylin, mounted in DPX and examined by light microscopy. Photographs were taken on Kodak TMAX 400 black and white film.

Table 4. Lectins used, their source, abbreviations and sugar specificities

Lectin	Abbreviation	Source	Specificities	Inhibitor
Ulex europaeus type I	UEA-I	Sigma[a]	α-1-Fucose	α-l-Fucose
Lotus tetragonolobus	LTA	Sigma	Fucα(1-2)Galβ (1-4)Fucα(1-3)GlcNAc	α-l-Fucose
Anguilla anguilla	AAA	EY[b]	α-l-Fucose	α-l-Fucose
Arum maculatum	AMA	Φ[c]	α-Mannose	α-Mannose
Allium sativum	ASA	Φ	α-Mannose	α-Mannose
Allium ursinum	AUA	Φ	α-Mannose	α-Mannose
Listera ovata	LOA	Φ	α-Mannose	α-Mannose
Narcissus pseudonarcissus	*NPA*	Φ	α-Mannose	α-Mannose
Galanthus nivalis	GNA	Φ	Manα3Man	α-Mannose
Hippeastrum hybrid	HHA	Φ	α-Mannose	α-Mannose
Psophocarpus tetragonolobus	PTA	Vector[d]	Fucα(1-2)Gal; d-GalNAc	d-GalNAc
Glycine max	SBA	Sigma	GalNAcα3GalNAc	d-GalNAc
Aegopodium podagraria	*APA*	Φ	d-GalNAc	d-GalNAc
Bryonia dioica	BDA	EY	d-GalNAc	d-GalNAc
Bauhinia purpurea	BPA	Φ	d-GalNAc	d-GalNAc
Dolichos biflorus	DBA	Sigma	d-GalNAc	d-GalNAc
Eranthis hyemalis	EHA	Φ	d-GalNAc	d-GalNAc
Iris reticulata	IRA	Φ	d-GalNAc	d-GalNAc
Viscum album	ML-II	Pfüller[e]	d-GalNAc	d-GalNAc
Sambucus nigra-II	SNA-II	Φ	Neu5Acα2,6Gal; GalNAc	d-GalNAc
Vicia villosa	VVA	Sigma	GalNAc; NeuAc	d-GalNAc
Helix pomatia	HPA	Sigma	d-GalNAc	d-GalNAc
Amaranthus caudatus	ACA	Φ	d-Gal	d-Gal
Griffonia simplicifolia I	GSA-I	Sigma	α-Gal; αGalNAc	d-Gal
Chelidonium majus	CMA	Φ	d-GlcNAc	GlcNAc
Euonymus europeus	EEA	Vector	Galα3,Fucα2,Galβ3; 4GlcNAc	GlcNAc
Colchicum autumnale	CAA	Φ	Complex	d-GalNAc

Fuc, fucose; Man, mannose; Gal, galactose; GlcNAc, *N*-acetylglucosamine; GalNAc, *N*-acetylgalactosamine; NeuAc, neuraminic acid (sialic acid).

[a] Sigma Chemical, Poole, Dorset, UK.

[b] EY Laboratories, San Mateo, Calif., USA.

[c] Own sources (for characterization, see Peumans and Van Damme 1994).

[d] Vector Laboratories, Peterborough, UK.

[e] Kindly provided by Dr. Uwe Pfüller.

3 Mucosal Responses to Diet and Microflora

3.1
Progress in Research of Intestinal Mucins

3.1.1
Dietary Components

The effects of dietary constituents on the morphology, cytokinetics and growth of the small as well as large intestine have attracted considerable attention in recent years. It is well known that the presence of both digestible and non-digestible components of the diet in the intestinal lumen exert a trophic effect on the mucosa. Initial response to luminal nutrients may be the modification of the crypt proliferation rate or cellular hypertrophy thus influencing the growth and functions of the gastrointestinal tract. In recent years, much attention has focused on the influence of dietary fibres on mucus secretion, since fibres are believed to protect the gastrointestinal mucosa from various diseases.

The feeding of low residue elemental diets (Goodlad and Wright 1983) and parenteral nutrition (Goodlad et al. 1992) have been shown to result in significant mucosal atrophy. There is also strong evidence that liquid fibre-free diets result in intestinal mucosal hypoplasia and reduced rate of cell proliferation (Sircar et al. 1983). This mucosal hypoplasia is reversed by consumption of a diet containing non-digestible fibre and leads to a hyperproliferative response in the large intestine (Jacobs and Lupton 1984). The mechanisms by which dietary fibre affect the intestinal mucosa are apparently complex and a number of hypotheses have emerged. Fibre was thought to exert trophic effects through its bulk either by mechanical distension (Dowling et al. 1969; Jacobs 1985) or by surface cell abrasion (Cassidy et al. 1981; Tasman-Jones et al. 1982). Recent studies, however, support the hypothesis that although the rate of epithelial cell proliferation in small intestine is regulated by the physico-chemical properties of diet (Pell et al. 1995), the growth of the colon is stimulated by the release of short chain fatty acids (SCFAs) from the fermentation of dietary polysaccharides by the bacterial flora (Sakata and Yajima 1984). This stimulatory response to fibre-rich diet in the colon is abolished in the germ-free animal (Goodlad et al. 1989), confirming that it is the products of bacterial degradation of undigested fibre, not fibre itself, that is responsible for colonocyte proliferation. Since the SCFAs exert proliferative effects on the colonocytes, a typical Western diet containing lower amounts of undigestible fibre has been implicated in the high occurrence of colonic disorders.

The role of dietary fat in regulating the intestinal cell proliferation is at present less concise and there have been conflicting reports on the trophic response to dietary lipids. Jacobs (1983) showed that a high fat diet produced a decrease in mucosal cell mass, whereas in the ileum only the unsaturated fat was associated with a reduction in mucosal weight and in cell hypoplasia. In contrast, Maxton et al. (1989) observed that consumption of long-chain triglycerides leads to increased mucosal mass and DNA content of the small intestine. Unsaturated fatty acids have also been shown to increase the epithelial cell proliferation in the small intestine (Jenkins and Thomson 1989). In the large intestine, however, the dietary fat is particularly influential in modulating the mucosal cell proliferation through the interactive effects with the dietary fibre (Lee et al. 1993). Differences in lipids derived from animal and vegetable sources may also influence the crypt cell proliferation (Pell et al. 1994). There is evidence that animals fed the high-fat diet show a reduction in the surface area of microvilli per enterocyte and a decrease in total proteins of the brush border membranes (Goda and Takase 1994). Although little is known about the mechanisms by which the luminal nutrients influence the mucosal proliferation, one hypothesis suggests that the workload of absorption imposed on the intestine is associated with stimulation of trophic intracellular messengers in the epithelial cells which provide signals for compensatory adaptation of the gut wall (Clarke 1977).

3.1.2 Host–Microflora Interactions in Intestinal Lumen

The skin and mucus membranes of all animals living under conventional conditions harbour 'normal microflora' and in mammals this flora is established only after birth. Compared to other organ systems, the gastrointestinal tract is host to a mass of microorganisms that colonise its epithelial surface. Within the last two decades, most of the work on ecology and composition of gut flora has been pioneered by Savage (reviews 1978, 1984, 1989), who introduced the concept of native indigenous and transient species of microflora. Most gut microorganisms are strict anaerobes, hence their reactions tend to be mainly reductive and hydrolytic, in contrast to those of mammalian tissue which are generally oxidative and conjugative (Coates and Walker 1992).

Over the past three decades, methods for culturing, sampling and identifying the gut flora have been developed. Light as well as scanning electron microscopic techniques have been used to study the microorganisms associated with the epithelial surfaces of the gastrointestinal tract (Savage 1980, 1983). These investigations have clearly shown the cross-sectional variation of microbial ecology at each level of the intestine; for example, the microbes are not only different in the large and small intestine but the flora that reside in the lumen differ from those at the epithelial surface (Gorbach 1975; Savage 1977). Along the epithelial surfaces, Gustafsson and Maunsbach (1971) demonstrated that the intestinal crypts harbour bacteria at various levels of depth in pure cultures.

The mechanisms of interaction between the microflora and intestinal mucosa are still poorly understood. From the literature available, however, it appears that the microbial challenge to the mucosal surface varies in location and this variation significantly alters the intestinal morphology and physiology.

The mucus layer plays a major role in the control of the microflora population. To remain in the mucus blanket, enteric bacteria would have to multiply at a rate exceeding that at which the mucus flows down the intestinal tract (Savage 1977), and in such a situation the organisms need to bind to the mucin glycoproteins. Although it is known that some bacteria survive on the nutrients derived from the mucin glycoproteins, in recent years research has been directed to test the hypothesis whether the nutrients derived from the diet influence the colonization of microflora. Most studies have been conducted in laboratory rodents. Starvation eliminates the anaerobic bacteria such as lactobacilli from the gut (Tannock and Savage 1974). Ducluzeau and Raibaud (1979) introduced a strain of bacteria in an isolator containing two groups of germ-free mice, one fed the commercial diet, the other a semi-synthetic diet. The establishment of bacteria in mice fed only the commercial diet indicated that indigestible dietary fibre is important in maintaining the population levels of the microflora associated with the epithelial surfaces. Many organisms found in the rat colon adhere to surfaces of digesta derived from the plant material (Harbers et al. 1980). The diet also influences which type of microbes colonise the intestines of ex-germ-free rats associated with conventional flora (Lhuillery et al. 1981).

3.1.3
The Human Microflora

To evaluate the effects of age and diets on the microflora, numerous studies of human gastrointestinal microbiodata have been conducted all over the world. Such studies analyse the faecal samples and in adults this microbiodata reflects the overall microbial population in the lumen of the large intestine (Moore et al. 1978). Data on direct sampling of the microflora from luminal content and mucosa at different sites in the gastrointestinal tract are scarce. Over 40 bacterial genera and 400 species have been isolated and identified from the faeces of individuals in several countries (Savage 1989). As an estimate of colonic content and the population levels of microflora, Savage (1977) indicated that as many as 100,000 billion bacteria live in the large bowel of humans. Despite changes in the diet, the population of bacterial species has been shown to be remarkably stable (Finegold et al. 1983). The findings of Norin et al. (1985) revealed that diet affects little the overall microbial composition, but can influence the biochemical activities of the bacteria present. The specific biochemical activities are complex and in many cases specific for particular floras under particular circumstances.

3.1.4
Relevance of Mucin Histochemical Studies

Present knowledge of the epithelial mucins synthesised by goblet cells is based on their intracellular localisation by some conventional and a few modern histochemical techniques. These methods have proved to be valuable and reliable in investigating the qualitative and quantitative differences in mucin-type glycoproteins between different parts of the gastrointestinal tract and at different levels of the intestinal crypts. In addition, they have been useful in identifying changes in the carbohydrate composition of mucins associated with colorectal diseases.

Histochemically, mucins can be divided into two main groups: neutral and acidic mucins (Drury and Wallington 1980). The periodic acid-Schiff (PAS) reaction is a general non-specific reaction for neutral mucins containing unsubstituted $^{\alpha}$-glycol groups (Filipe and Lake 1983). The chemical basis of the PAS technique is well established, whereby dialdehydes released from vicinal glycols by periodate oxidation form, with Schiff reagent, a magenta-coloured product in the tissue and have the additional advantage of being quantitative (Pearse 1985). Acidic mucins are either sialomucins or sulphomucins and the relative contribution of sialic acid and ester sulphate moieties to the acidic characteristics of mucins can be demonstrated by Alcian blue (AB) which at buffered-controlled pH distinguishes between the two groups. AB at pH 2.5 will ionize and stain both carboxyl groups of sialic acids and the sulphate ester groups (Drury and Wallington 1980). However, at pH 1.0 sulphate esters are selectively stained with AB while the carboxyl groups are undissociated (Lev and Spicer 1964). A combined procedure involving the PAS and AB at pH 2.5 staining methods can also be applied to simultaneously visualise, and distinguish between, the neutral and acidic mucins or a mixture of both by contrasting colours on the single section (Mowry 1963). To obtain valid results with Alcian blue histochemical methods, a dot blot assay technique to validate the suitability of this dye to stain carboxylated acidic and sulphated mucins was developed by Schumacher and Adam (1992).

The introduction of lectins, carbohydrate specific glycoproteins of plant or animal origin, to the mucin histochemistry has allowed the detection of terminal carbohydrate residues of mucin glycoconjugates, this specific type of analysis was impossible with the conventional methods (Spicer and Schulte 1992; Danguay et al. 1994). Recognition of lectins can be traced back to the last century when Stillmark (1888) described the phenomenon of haemagglutination by ricin derived from castor beans (Bird 1959). Although their biological function in plants and animals has not definitely been clarified, lectins have been widely used as probes for carbohydrate analysis in cell biology and pathology. Conjugated to a label visible by light microscopy, lectins serve as histochemical markers to examine selectively glycoconjugates to which the lectin binds. The interaction of glycoproteins with lectins may be likened to the antigen-antibody reaction and can be inhibited by simple sugars (Sharon and Lis 1972). Most lectins possess two or more carbohydrate binding sites and the affinity with which a lectin binds to a sugar is influenced by the position of the glycan residue within an oligosaccharide chain. At the cellular level, lectins detect cytoplasmic components rich in glycoconjugates such as cell surface membrane or secretory granules, lysosomes and the Golgi apparatus involved in synthesis, storage or turnover of complex carbohydrates (Damjanov 1987).

Because aspects of mucosal defence mechanisms are vital for understanding disease, studies of mucin synthesis and secretion constitute an important area of ongoing research. With the application of newer investigation techniques in recent years, the major advances in mucin research have been from in vivo studies on gene regulation with increasing availability of mucin cDNAs. With the diversity of mucin species and with the degree of cellular localisation by histochemical methods that have been shown within the intestinal tract, many questions arise. We need, for example, to know how does production of different mucins adapt to the changes in the luminal environment and what are the unique functional benefits to the epithelial surface of the individual mucins? How are the various mucins related and what factors initiate the synthesis and secretion of mucins? Do microflora–epithelial interactions affect the composition and turnover of different types of mucins? The investigations conducted for the work and methodology desribed in this monograph have attempted to answers these questions.

Epidemiological evidence points towards diet as a risk factor for the development of gastrointestinal diseases such as colorectal cancer and to some extent this is supported by animal studies. As the evidence has accumulated in the literature of trophic effects of dietary fibre on morphology and cell proliferation rate of the gastrointestinal epithelium, the need to understand the response of interactions between the diet and microflora on goblet cells mucins becomes imperative. Because mucin is an important nutrient source for many intestinal organisms, the impact of dietary constituents, particularly the fibre, may influence the microbial population present. The work described in this monograph attempts to clarify the intraluminal impact of food and bacteria in an experimental model of a germ-free rat fed on different diets and associated with either conventional flora or human flora.

The data available from literature have established that in gastrointestinal diseases the structural changes in goblet cell mucins are accompanied by alterations in their carbohydrate sequences and the introduction of lectins to the histochemical analysis of glycoprotein structures has improved the detection of mucin oligosaccharide structure. It is generally accepted that there is a sequential program for goblet cell mucin glycosylation, in which carbohydrate residues such as Gal, GalNAc and Fuc may be added terminally. Numerous studies have indicated that the state of glycosylation depends not only on the stage of differentiation and maturation of goblet cells along the crypt-villus axis but also along the oral aboral axis of the intestinal tract. Unfortunately, the information on the effects of diet or the presence/absence of bacteria on glycosylation is scarce. Since a critical step in the bacterial colonisation of the gut is the binding of the organism to the epithelium through its surface adhesins, the diet may alter the bacterial ecology of the intestinal tract by changing the glycoconjugate expression of epithelial mucins. The results derived from the staining pattern of mucins by a comprehensive panel of lectins would also constitute a starting point for systematically examining the effects of diet and microflora on the activity of carbohydrate-modifying enzymes such as glycosyltransferases and glycosidases.

3.1.5 Morphological Effects

Most of our knowledge of dietary effects on intestinal morphology is derived from animal studies and practicalities of extrapolations from animal model systems to human conditions must be critically evaluated. Multiple mechanisms are involved in the process of epithelial cell renewal in the crypts and villi thus changing the morphology of the intestinal luminal surface. Assuming that enterocytes and mucous cells differentiate from a common stem cell at the base of the crypts and undergo transit divisions along the villus column, then changes in either enterocyte or goblet cell function could be responsible for changes in villus and crypt dimensions.

One of the factors which may alter the appearance of villi is the nutritional status of the individual. Dietary manipulation can modify the microvillar structure of the epithelial cells thus modifying the absorptive surface of the intestinal mucosa. The shape of intestinal villi in humans is subject to alterations with variations in dietary constituents (Creamer 1974). Similarly, in rats weaned on a diet containing fibre there is a change in villus structure from finger-like villi to broad-leafed, long-ridged villi of the mature animal (Tasman-Jones et al. 1982). Morphological findings with feeding of pectin and cellulose (Brown et al. 1979) suggest that the alterations in gut physiology evoked by chronic feeding of these particular fibres may be associated with structural changes in the villus. Changes in villus morphology have also been associated with effects of fibre components on intestinal microflora (Savage 1978) and bile-acid metabolism (Chadurek et al. 1979). Marked villus enlargement has been shown in the ileum receiving the pancreatic secretion (Altmann 1971) and there is evidence that this secretion is stimulated by diet containing wheat bran (Schneeman et al. 1982).

It is well known that the dietary components alter the villus height and crypt depth thus modifying the surface area available for absorption, but the functional correlation of these changes with nutrient digestion and transport have been less well documented. However, the villus height and crypt depth still remain useful indicators of the size of the absorptive and proliferative compartments, respectively, in the intestinal mucosa.

3.2 Results and Discussion

3.2.1 Histochemical Studies

The histochemical staining pattern of carbohydrate-containing surface mucus, surface goblet cells and upper and lower crypt goblet cells is summarised in Tables 5 and 6.

3.2.1.1 General

Neutral and acid mucins were observed in the surface mucus and in all the goblet cells examined in both the small and large intestines. Acidic mucins were predominant in the lower crypt goblet cells of the large intestine. Sulphomucins, although present at other sites, were predominant in the upper crypt goblet cells of both the small and large intestine.

3.2.1.2 Effects of Diet

The effects of diet were assessed by comparing the staining characteristics in germ-free (GF; Fig. 3a,b) and conventional-fed CV rats (Fig. 3c,d) receiving either the commercial or purified diets (see Table 5). Although minor changes were seen throughout

Table 5. Effects of diet on histochemical characteristics of intestinal mucins in germ-free (*GF*) and conventional (*CV*) rats

Intestine and cell type	GF-C PAS	AB2.5/ PAS	AB 1.0	GF-P PAS	AB 2.5/ PAS	AB 1.0
Small intestine						
Surface mucus	+	+Purple	(+)	++	+Pink ++DP	(+)
Surface goblet cells	++	+++DP	+	+++	++DP	++
Upper crypt goblet cells	+++	+++DP	++	++	++Purple	++
Lower crypt goblet cells	+	++DP	0	++	++DP	+
Large intestine						
Surface mucus	(+)	+Purple	(+)	+	++Purple	(+)
Surface goblet cells	+	++Pink	0	++	+Pink ++Purple	(+)
Upper crypt goblet cells	+	+Pink	+++	+++	++DP	++
Lower crypt goblet cells	++	++Blue	(+)	+	+++Blue	(+)
	CV-C			CV-P		
Small intestine						
Surface mucus	++	+Purple	0	(+)	+Pink ++Purple	0
Surface goblet cells	+++	++DP	(+)	++	++DP +Purple	++
Upper crypt goblet cells	+++	+Purple	++	++	+Pink ++Purple	+
Lower crypt goblet cells	++	++DP	(+)	+	++Purple ++DP	(+)
Large intestine						
Surface mucus	+	+Purple	(+)	0	+Purple	0
Surface goblet cells	+++	+Pink +Purple	(+)	+	++Purple +Pink	++
Upper crypt goblet cells	+++	++DP	++	++	++Purple ++Pink	++
Lower crypt goblet cells	++	++Blue	+	(+)	++Blue +Purple	(+)

Intensity of reaction: 0 no reactivity; (+) weak reactivity; + moderate reactivity; ++ strong reactivity; +++ intense reactivity.

AB, Alcian blue; C, commercial diet; CV, conventional; DP, deep purple; GF, germ-free; P, purified diet; PAS, periodic-acid Schiff.

Table 6. Effects of microflora on histochemical characteristics of intestinal mucins in conventional (*CV*) and human flora-associated (*HFA*) rats in groups fed a purified diet

Intestine and cell type	GF-P			CV-P			HFA-P		
	PAS	AB2.5/ PAS	AB1.0	PAS	AB2.5/ PAS	AB1.0	PAS	AB2.5/ PAS	AB 1.0
Small intestine									
Surface mucus	++	+Pink k ++DP	+	+	+Pink ++Purple	0	++	+Pink +DP	0
Surfaceg goblet cells	+++	+DP	++	++	++DP +Purple	++	++	++DP	++
Upper crypt goblet cells	++	++Purple	++	++	+Pink ++Purple	+	+++	+Purple	+++
Lower crypt goblet cells	++	++DP	+	+	++Purple ++DP	(+)	(+)	+++DP	(+)
Large intestine									
Surface mucus	+	++Purple	(+)	0	++Purple	0	0	+Purple	++
Surface goblet cells	++	+Pink ++Purple	(+)	+	+Purple +Pink	++	+	++Pink +Purple	+
Upper crypt goblet cells	+++	++DP	++	++	++Purple ++Pink	++	++	++DP	+++
Lower crypt goblet cells	+	+++Blue	(+)	(+)	++Blue +Purple	(+)	(+)	+++Blue	(+)

Intensity of reaction: 0 no reactivity; (+) weak reactivity; + moderate reactivity; ++ strong reactivity; +++ intense reactivity.
AB, Alcian blue; C, commercial diet; CV, conventional; DP, deep purple; GF, germ-free; HFA, human flora-associated; P, purified diet; PAS, periodic-acid Schiff.

Fig. 3 a–d. Effects of diet on goblet cells mucins. **a** Small intestinal villi from a germ-free rat fed on a commercial diet stained by AB-PAS technique. Surface goblet cells are stained *deep purple* indicating that they contain neutral and acidic carboxylated mucins. ×330. **b** Large intestinal crypts from a germ-free rat on a commercial diet stained by AB-PAS technique. The surface and crypt goblet cells are predominantly stained *pink* (*arrows*). ×330. **c** Large intestine of a conventional rat fed on a purified diet stained with AB-PAS procedure. Goblet cells in the upper crypt contain neutral mucins. ×660. **d** Large intestinal crypts of a conventional rat fed on a purified diet stained with AB at pH 1.0. The goblet cells containing sulphated mucins are shown (*arrow*). ×330

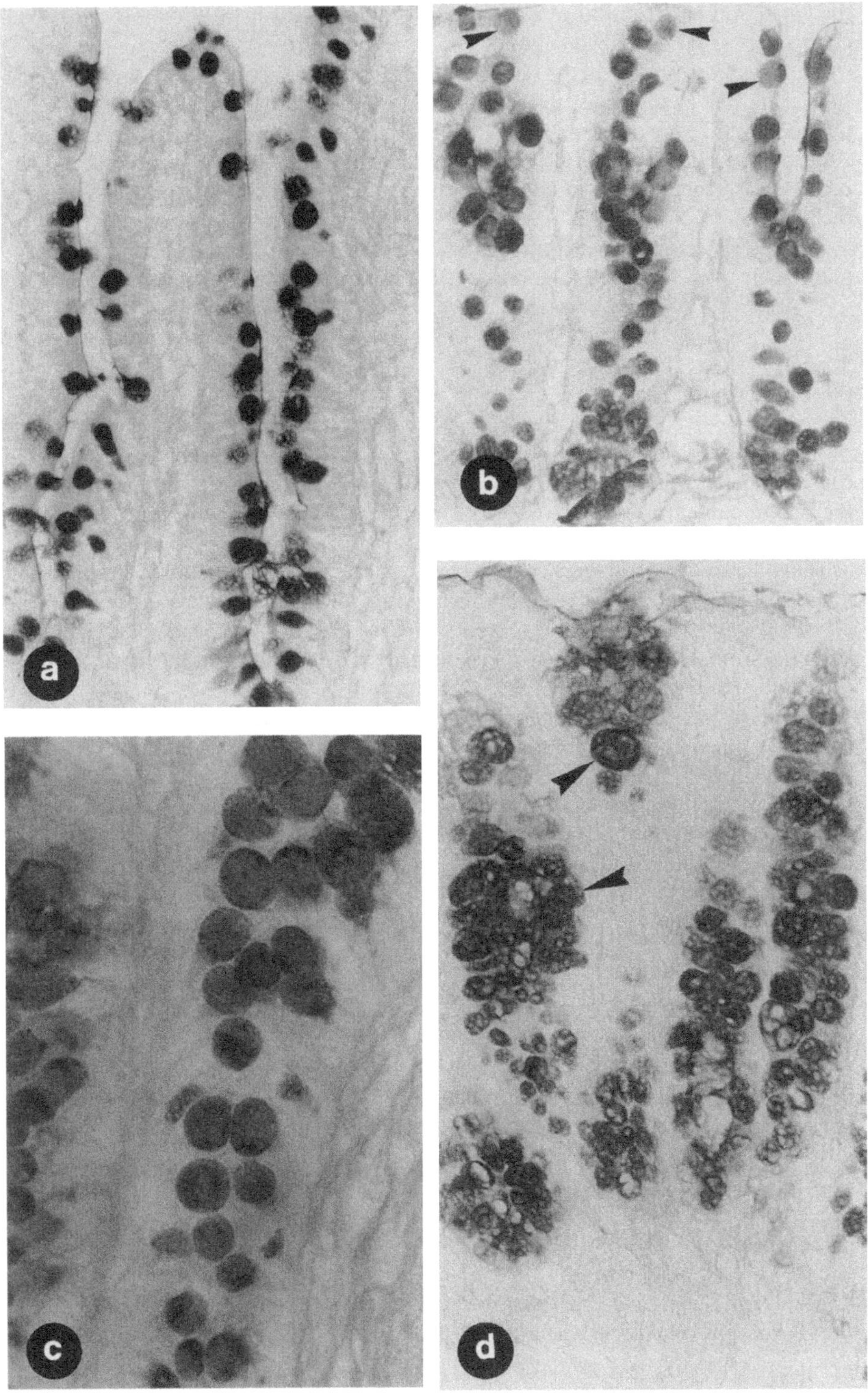
a
b
c
d

the small and large intestines as a result of the different diets, the most consistent changes are those described below.

Changes were present in the neutral mucins of the surface mucus and of the goblet cells examined in the small and large intestines of both GF and CV rats. In GF rats neutral mucins were more intensely stained at these sites in animals fed a purified diet when compared to animals fed a commercial diet. However, in conventional rats the reverse was true, i.e. less intense staining of neutral mucins occurs following the purified diet.

GF and CV animals fed a purified diet also showed increased staining for sulphomucins in the surface goblet cells of both the small and large intestines.

3.2.1.3
Effects of Microflora

The effects of normal rat flora on mucin content were assessed by comparing the conventional (CV) rats with the germ-free (GF) rats and the effects of human flora on the mucin content were assessed by comparing the HFA rats with the GF rats (see Table 6).

When GF and CV rats were compared to indicate the effects of normal rat flora on mucin content, there was a reduction in staining for neutral mucins observed in the surface mucus and many of the goblet cells examined; this was particularly so in the large intestine. There was also reduced staining for sulphomucins in the small intestine and in the surface mucus of the large intestine.

The above-mentioned changes were present irrespective of whether the animals were fed a purified or a commercial diet.

When GF and HFA rats were compared to indicate the effects of a human flora on mucin content, there was some reduction in staining for neutral mucins in the small intestine but a more marked reduction was observed in the large intestine. These changes were similar to those seen with the rat flora. There was little change in the acid mucin content of the goblet cells of the small intestine whereas in the large intestine the amounts present were reduced as a result of the presence of human flora.

3.2.2
Quantitative Morphology

The data on the effects of diet on the villus and crypt lengths and the villus/crypt ratio are shown in Fig. 4a,b and the effects of flora in Fig. 5a–c.

3.2.2.1
Effects of Diet

There were no significant changes observed in the villus lengths as a result of diet in the GF animals (see Fig. 4a). The GF rats fed a purified diet had shorter crypt lengths

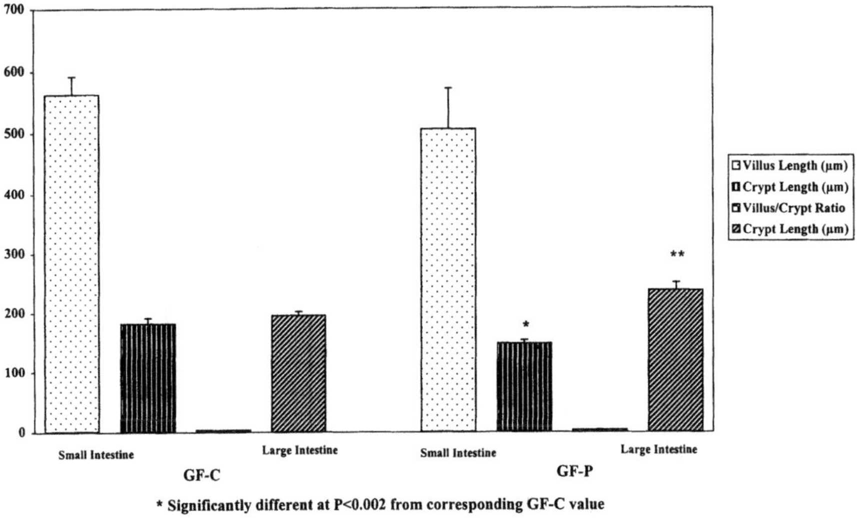

a

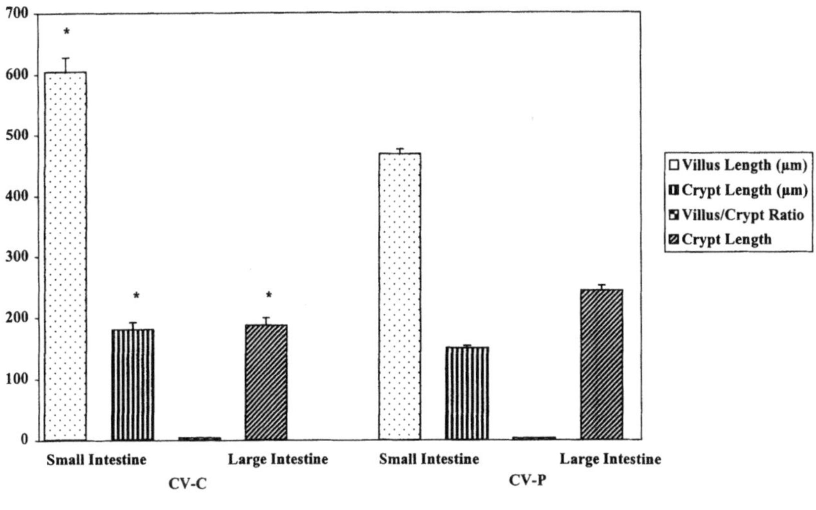

b

Fig. 4. a Effects of diet on villus and crypt lengths in germ-free rats (means±SEM). *Significantly different at $P<0.002$ from corresponding GF-C value. **Significantly different at $P<0.05$ from corresponding GF-C value. **b** Effects of diet on villus and crypt lengths in conventional rats (means±SEM). *Significantly different at $P<0.001$ from corresponding GV-P value

in the small intestine but longer crypts in the large intestine. However, in the CV animals the villus and crypt length in the small intestine was significantly shorter in the animals fed a purified diet whereas the crypt length in the large intestine was significantly increased (see Fig. 4b)

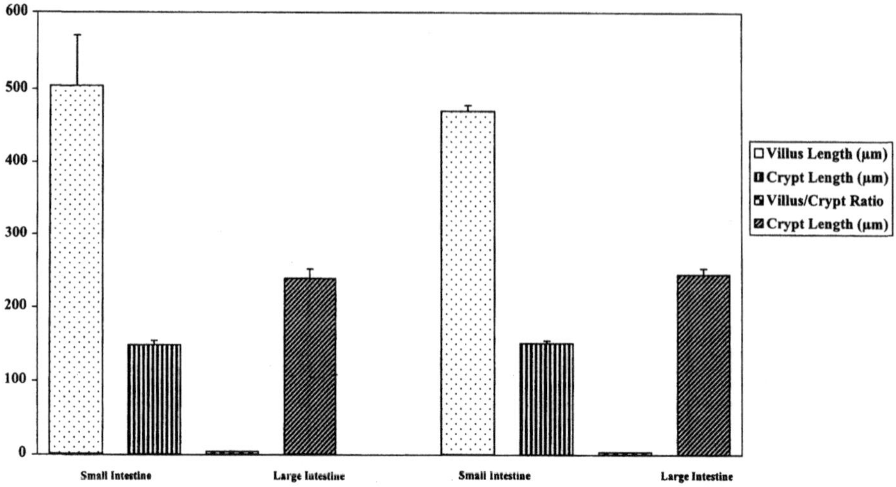

a GF-P CV-P

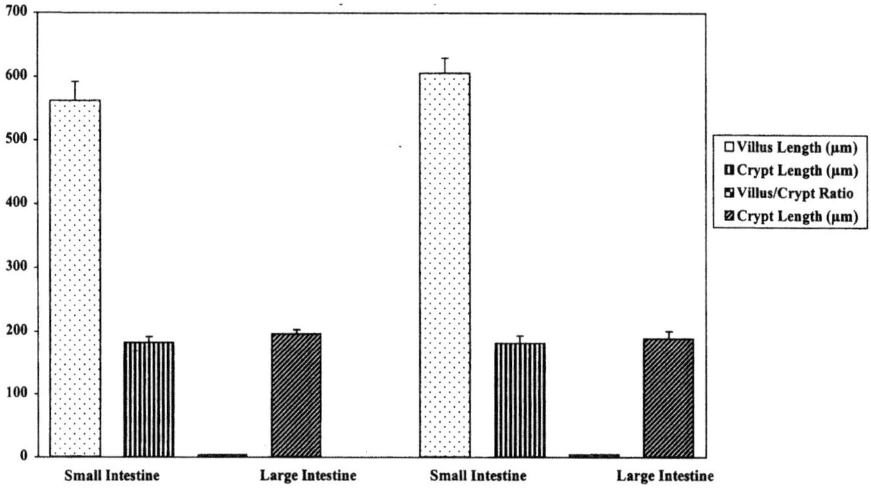

b GF-C CV-C

Fig. 5. **a** Effects of conventional flora on crypt and villus lengths in rats fed a purified diet (means±SEM). **b** Effects of conventional flora on crypt and villus lengths in rats fed a commercial diet (means±SEM). **c** Effects of human flora on crypt and villus lengths in rats fed a purified diet (means±SEM).

* Significantly different at $P<0.001$ from corresponding HFA value

3.2.2.2
Effects of Microflora

The presence of normal rat flora produced no significant differences in the villus and/or crypt length of the small or large intestine (see Fig. 5a,b). This lack of effect was apparent irrespective of which diet the animals had received.

When HFA rats were compared with GF rats to test the effects of human flora on villus and crypt lengths, significant changes were observed (see Fig. 5c). In the presence of human flora the crypt length in the small and large intestine was significantly lower than that seen in the GF animals.

3.2.2.3
Discussion

The results from this experiment have confirmed previous histochemical observations (Reid et al. 1974; Sheahan and Jervis 1976; Altmann 1983) that the composition of secretory glycoproteins in the intestinal mucosa of rats differs with the region and with the cell type. These secretory glycoproteins are also influenced by diet and by the presence or absence of a microbial flora in the intestinal lumen.

In the conventionally reared animals it is the neutral mucosubstances that predominate in the small intestine and acid mucosubstances in the large intestine. Furthermore, the goblet cells in the upper part of the large intestinal crypts differed from those in the lower part in that they contained Alcian blue pH 1.0 reactive sulphated mucins. Although similar results have been described in previous histochemical investigations (Sheahan and Jervis 1976; Shamsuddin and Trump 1981; Thomopoulos et al. 1983; Tsuyama et al. 1983; Park et al. 1987), the significance of the differences in mucin distribution patterns throughout the intestinal tract and functions of various classes of intestinal mucins is still not well understood.

The influence of diet on the composition of intestinal mucins was explored in the GF and CV rats fed one of two diets. The first diet contained finely powdered, purified ingredients, including cellulose as a source of fibre. The commercial diet contained a more coarsely ground diet of natural ingredients with crude fibre of mainly cereal origin.

The results of this study clearly showed that the purified diet led to an increase in neutral mucins in the small and large intestines of the GF animals but to a decrease of these mucins in the CV animals. An increase in sulphated mucins was also observed in the small and large intestines of animals fed the purified diet. Although the mechanisms by which diet modifies mucin composition in germ-free animals are not clearly understood, it is possible that the purified diet in the GF animals is more difficult to break down than the commercial diet whereas the bacterial mucinase may assist the breakdown in the CV animals.

The histochemical studies on responses to dietary fibres have so far been limited to jejunal goblet cells of pigs which indicate an increase in staining intensity of sialylated acidic mucins when animals are fed a diet rich in bran, pectin or gum (Moré et al. 1987). There are other studies which report that specific dietary fibres such as wheat bran may increase the secretory activity of goblet cells in rodents (Cassidy et al. 1982; Schneeman et al. 1982; Vahouny et al. 1985).

A comparison of the GF rats with their counterparts harbouring indigenous flora showed that, independently of diet, the intestinal tract responded to intraluminal contamination by reducing the amount of neutral mucins in the goblet cells of the both the small and large intestine and increasing the amount of acid mucins in the small intestine and in the surface mucus and goblet cells in the large intestine. It is well known that the amount of mucus in the intestinal lumen of GF rats is greater than that in their CV counterparts, because of the degradative effect of bacterial mucinase (Heneghan 1984).

However, in the present study where staining intensity was used as the parameter for identifying mucins, it is not possible to draw a direct analogy with the amount of mucus present. The degree of glycosylation of mucus glycoproteins and hydration of secreted mucus may have been partly responsible for the different staining intensities seen in the animal used in the present study.

Changes in carbohydrate composition of intestinal mucins have been known to occur during development (Brackett and Townsend 1980; Sharma and Schumacher 1992) and during the migration of crypt cells toward the epithelial surface (Park et al. 1987). It is well known that the migration of epithelial cells along the villi is faster in CV compared with GF animals (Heneghan 1979). In this study, in the presence of a conventional flora goblet cell mucins in the large intestine became more sulphated along the crypt villus axis. This is consistent with studies in the neonatal mouse, where similar effects of the conventional flora on the production of mucins have been described (Hill et al. 1990). According to Heneghan (1984), who reviewed the gastrointestinal tract of gnotobiotes, the villus/crypt ratio is higher in GF animals as compared to conventional animals. This is in contrast to the present findings which indicate comparable lengths of villus column and crypts in germ-free animals and conventional animals fed on either the purified or the commercial diet. Similar results to those described in this work have been observed in studies by Ishikawa et al. (1986) in the upper region of small intestine. In the present study, when the morphological effects of the two diets were compared in germ-free animals, the most remarkable alterations were seen in the crypt lengths of small and large intestines. Crypt size was decreased in the small intestine of rats fed a purified diet, and although the villus height did not change significantly, it was clear that the increased crypt length in the large intestine observed in this group of animals reflected an increased amount of neutral mucins. The finding of no changes in villus length is also in contrast to the results of a morphological study on conventionally fed rats by Sigleo et al. (1984). This discrepancy implies that there may have been an interaction between the diet and microbial activity (Savage 1978) in their studies.

Although the mechanisms responsible for structural and chemical changes in the intestinal mucosa are unknown, the results of this study indicate that the presence of a microbial flora and the nature of the diet can determine the shape and mucus secretion of the intestinal epithelium and its supporting stroma. Since the thickness of the surface mucus and the chemical composition of surface mucins is linked to a dynamic equilibrium between the continuous secretion of mucin from the goblet cells and its degradation within the intestinal lumen, it is appropriate to assess the composition of the surface mucus in response to intraluminal stimuli of different diets and different species of gut flora. The evaluation of surface mucus is prone to artifacts of fixation and should be considered with caution. However, it appears reasonable to assume that the comparison between staining intensities of surface mucus in all animal groups, if all samples are processed at the same time and analysed by standardised histological methods, are valid. The effects of the human flora on the gut structure and mucus composition were in many ways similar to those of the indigenous rat flora. However, there were some differences, for instance in the composition of the surface and goblet cell mucus and the length of the large intestinal crypt cells, which merit more detailed investigation. Although they may reflect real differences in response to the two types of flora, it is also possible that when a human flora is inoculated in the germ-free animal, its establishment and subsequent interaction with the intestinal tract may induce changes not seen with an indigenous flora where colonisation is not subject to experimental manipulation.

The findings of altered mucosal morphology and mucin biosynthesis with particular dietary patterns in the presence of an indigenous or a human microbial flora described in this work strongly emphasises the use of the rat as a model system for pathogenesis of intestinal disease. The results from human flora rats are preliminary and more detailed studies are needed to ascertain whether the HFA rat may be a more appropriate model for man than the ordinary conventional rat. Further studies of changes in mucin composition by various intraluminal stimuli will help in understanding the mechanisms of intestinal disorders and in developing probes for detection of bowel disease. To further elucidate the interaction between the mucin secretion and intestinal luminal components in the animal model, more detailed analysis of goblet cell glycoproteins were carried out using an image processor.

3.2.3
Morphometric Analysis of Intestinal Mucins

The analysis of the synthesis of mucins is a cumbersome task since a distinctive heterogeneity of neutral and acid glycoprotein secretion exists not only between the goblet cells of the small and large intestine, but also between adjacent goblet cells at both sites (Lamont 1992). Furthermore, the sulphated mucins are confined to the deep crypt secretory cells in the colon (Park et al. 1987) stressing the need to analyse these mucins at the histochemical level where adjacent goblet cells can be studied individually. The need for in situ analysis is further emphasised by the fact that changes in intestinal mucins are directly linked to a dynamic equilibrium between their biosynthesis by goblet cells and their degradation within the lumen (Allen 1981). It is pertinent, therefore, to quantitate the effects of dietary manipulations and bacterial flora on the synthetic and secretory levels of neutral, acid and sulphated glycoproteins in mucin-containing cells. Studies described in Sects. 3.2.1 and 3.2.2 have shown by using germ-free and conventional rats that the gut microflora, as well as the dietary composition, can influence both the mucosal architecture and the secretory pattern of the intestinal mucins. The aim of the study described here was to define more clearly the effects of diet and microflora by the use of an image analysis system.

An image analysis system was used to evaluate the amount of intracellular glycoproteins in the mucin-containing cells of defined parts of the jejunum and colon. The jejunum was chosen for analysis because diet has been implicated in the stimulatory effects on the secretory activity of the jejunal goblet cells (Vahouny et al. 1985) and the colon because of the effects of the mucin-degrading flora in altering of the normal mucin pattern (Carlstedt-Duke 1989).

3.2.3.1
Effects of Diet

The influence of the commercial or of the purified diet on the number of mucin-containing cells, the volume densities and the staining densities of their mucin granules are shown in Fig. 6a–d.

3.2.3.1.1
Number of Mucin-Containing Cells

The number of mucin-containing cells was similar in the small intestinal villi and large intestinal crypts of germ-free rats and in the large intestinal crypts of conventional rats fed on two different diets. The number of acidic mucin-containing cells in villi of conventional rats was significantly greater in the group fed on the commercial diet (see Fig. 6a).

3.2.3.1.2
Volume Densities of Mucin Granules

The volume density of granules containing sulphomucins in the small intestine was significantly less in the germ-free rats fed on the commercial diet when compared to those fed on the purified diet (see Fig. 6b). In conventional rats fed on the commercial diet the volume densities of mucin granules containing neutral mucins and sulphomucins in the small intestine was significantly less (see Fig. 6b). No differences in volume densities of mucin granules of the large intestine were observed between the two germ-free and two conventional dietary groups.

3.2.3.1.3
Staining Density of Mucin Granules

In the small intestine, the staining densities of neutral and acidic mucin granules were significantly less in cells of germ-free rats fed the commercial diet but that of sulphomucin granules was increased when compared with their counterparts given the purified diet (see Fig. 6c). The staining density of neutral mucin granules in the cells of the small intestine of conventional rats fed on the commercial diet was significantly raised (see Fig. 6d). In the large intestine of germ-free rats fed the commercial diet the mean staining densities of neutral, acidic and sulphated mucin granules in the crypt secretory cells were significantly higher than those seen in germ-free rats fed the purified diet (see Fig. 6c). However, in the large intestine of conventional animals, the mean staining density of neutral and sulphomucin granules was lower in the group fed on a commercial diet (see Fig. 6d).

3.2.3.2
Effects of Microflora

The effects of conventional microflora and of the interactions of the conventional flora with the diet on the secretory pattern of intestinal mucins was assessed by comparing the germ-free and conventional rats receiving the commercial diet and between germ-free and conventional rats receiving the purifed diet. The effects of human flora were assessed by comparing germ-free and HFA rats receiving the puri-

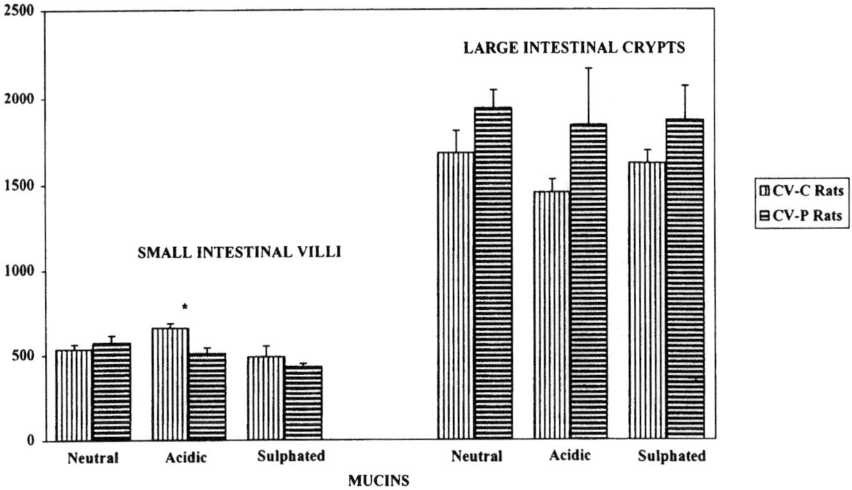

a * Significantly different at P<0.01 from corresponding value for conventional rats fed a purified diet

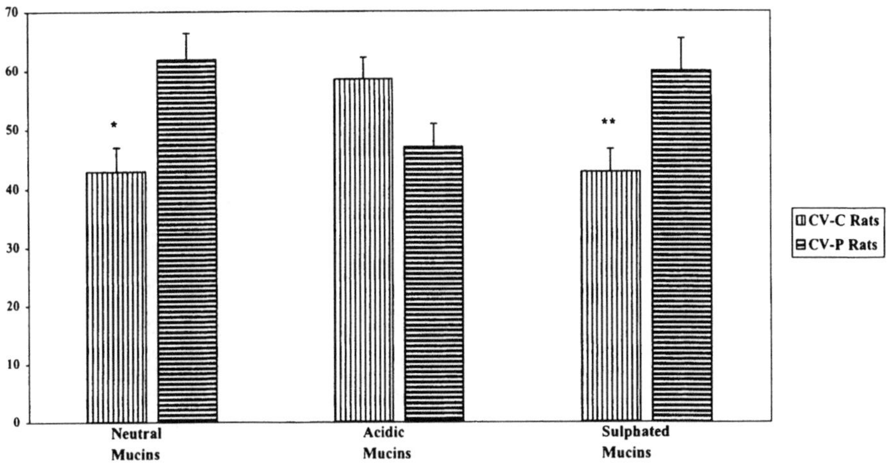

* Significantly different at P=0.002 from corresponding value for conventional rats fed a purified diet

b ** Significantly different at P<0.02 from corresponding value for conventional rats fed a purified diet

Fig. 6. a Effects of diet on the number of mucin-containing cells/mm^2 in the small and large intestines of conventional rats (values are means±SEM). *Significantly different at $P<0.01$ from corresponding value for conventional rats fed a purified diet. **b** Effects of diet on volume density of mucin granules (μm2) in the small intestines of conventional rats (values are means±SEM). *Significantly different at $P<0.002$ from corresponding value for conventional rats fed a purified diet. **Significantly different at $P<0.02$ from corresponding value for conventional rats fed a purified diet. **c** Effects of diet on staining density of mucin granules in the small and large intestines of germ-free rats (values are means±SEM). *Significantly different at $P<0.0001$ from corresponding value for germ-free rats fed a purified diet. **Significantly different at $P<0.001$ from corresponding value for germ-free rats fed a purified diet. **d** Effects of diet on staining density of mucin granules in the small and large intestines of conventional rats (values are means±SEM). *Significantly different at $P<0.0001$ from corresponding value for conventional rats fed a purified diet. **Significantly different at $P<0.02$ from corresponding value for conventional rats fed a purified diet. ***Significantly different at $P<0.0001$ from corresponding value for conventional rats fed a commercial diet

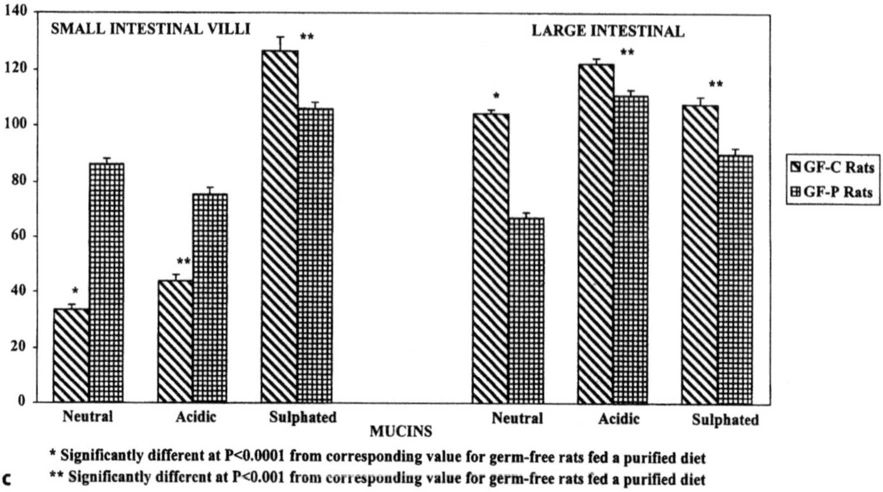

c

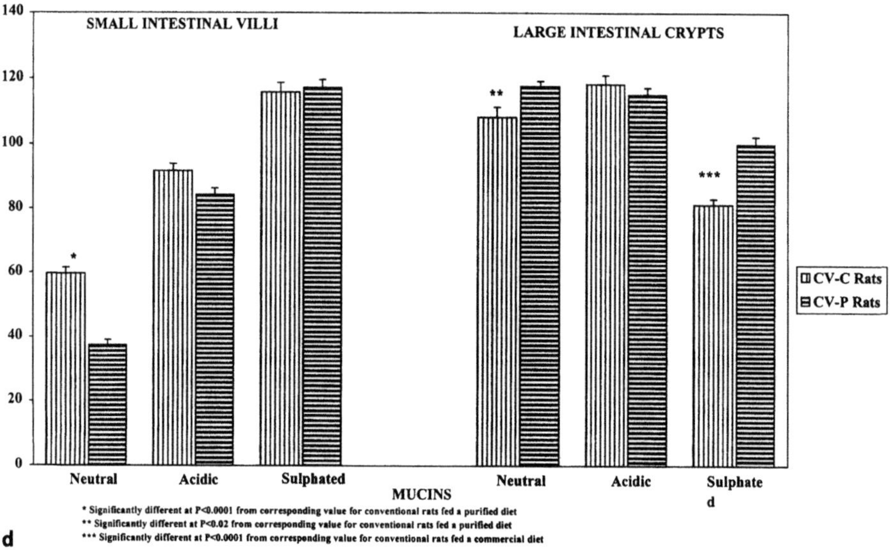

d

c Effects of diet on staining density of mucin granules in the small and large intestines of germ-free rats (values are means±SEM). *Significantly different at $P<0.0001$ from corresponding value for germ-free rats fed a purified diet. **Significantly different at $P<0.001$ from corresponding value for germ-free rats fed a purified diet. **d** Effects of diet on staining density of mucin granules in the small and large intestines of conventional rats (values are means±SEM). *Significantly different at $P<0.0001$ from corresponding value for conventional rats fed a purified diet. **Significantly different at $P<0.02$ from corresponding value for conventional rats fed a purified diet. ***Significantly different at $P<0.0001$ from corresponding value for conventional rats fed a commercial diet

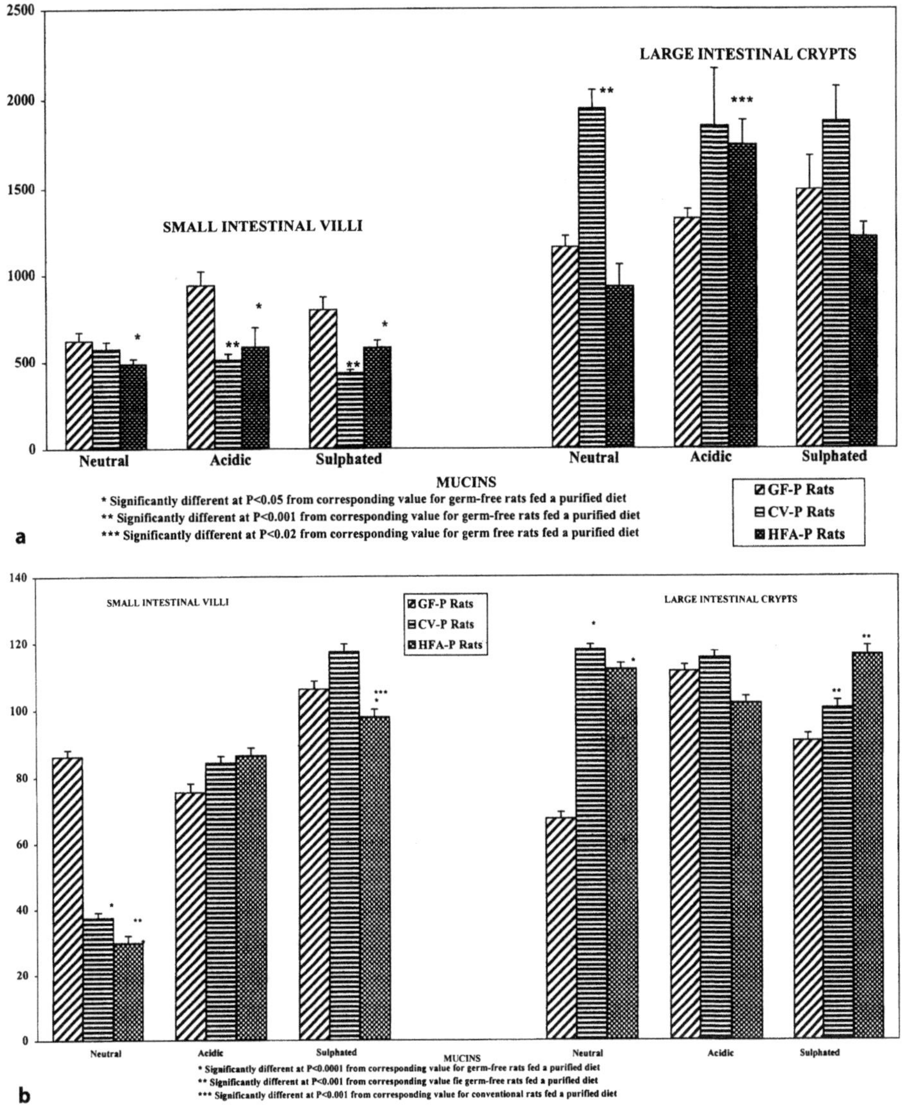

Fig. 7. **a** Effects of microflora on number of mucin-containing cells/mm2 in the small and large intestine of the rat fed a purified diet (values are means±SEM). *Significantly different at $P<0.05$ from corresponding value for germ-free rats fed a purified diet. **Significantly different at $P<0.001$ from corresponding value for germ-free rats fed a purified diet. ***Significantly different at $P<0.02$ from corresponding value for germ-free rats fed a purified diet. **b** Effects of microflora on staining density of mucin granules in the small and large intestine of the rat fed a purified diet (values are means±SEM). *Significantly different at $P<0.0001$ from corresponding value for germ-free rats fed a purified diet. **Significantly different at $P<0.001$ from corresponding value for germ-free rats fed a purified diet. ***Significantly different at $P<0.001$ from corresponding value for conventional rats fed a purified diet

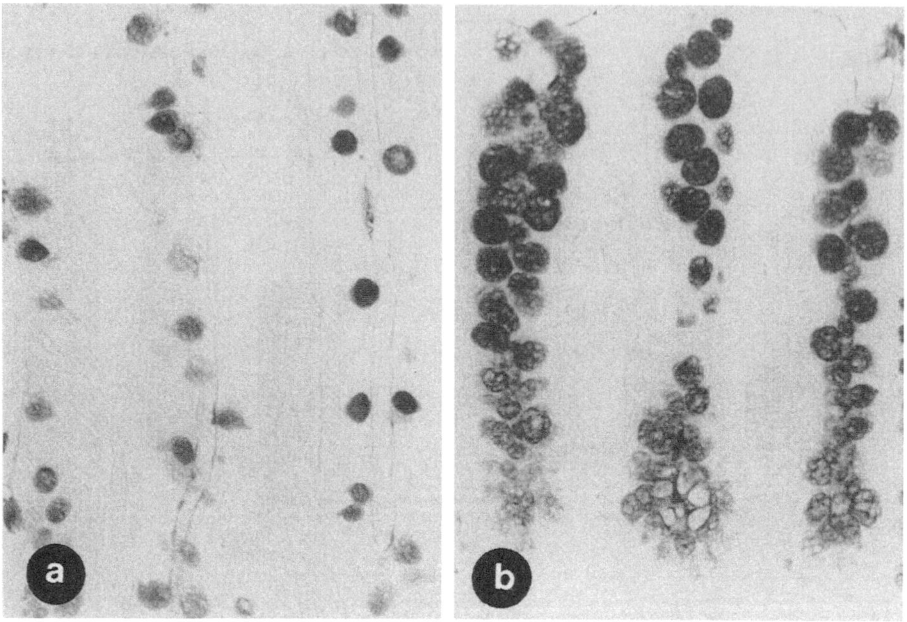

Fig. 8. Mucin-containing cells of the small intestinal villi from a conventional rat fed the purified diet stained by AB at pH 2.5 (**a**) and of the large intestinal crypts from an HFA rat fed the purified diet stained by AB at pH 1.0 (**b**). Note the differences in distribution, volume density and staining density of carboxylated acidic and sulphomucins at two sites in two groups. ×265

fied diet. In groups fed on the purified diet, the effects of the two different microflora were assessed by comparing the conventional and HFA rats. The results are shown in Fig. 7a,b. Figure 8a shows mucin-containing cells in the small intestine from a CV-P rat stained by AB at pH 2.5 and Fig. 8b in the large intestine of an HFA-P rat stained by AB at pH 1.0.

3.2.3.2.1
Number of Mucin-Containing Cells

In both the small and large intestine, the numbers of neutral, acidic and sulphomucin-containing cells were similar in germ-free and conventional rats fed the commercial diet (Fig. 7a). However, in rats fed on the purified diet there were significantly less cells containing acidic and sulphomucins in the small intestine and more cells containing neutral mucins in the large intestine of the conventional animals as compared to those of the corresponding germ-free group. A significantly lower number of all three types of mucin-containing cells was found in the villi of rats with a human flora when compared with the corresponding germ-free rats.

No significant differences in the numbers of neutral and sulphomucin-containing crypt secretory cells were found in the large intestine but the number of acidic mucin-containing cells was significantly more in HFA rats compared with their germ-free counterparts. In rats with a human flora, the numbers of sulphomucin-containing cells were significantly more in the small intestine but significantly less in the large intestine than those with a conventional rat flora. Cells containing neutral mucins were significantly fewer in number in crypts of the large intestine of HFA rats than in those of the conventional rats.

3.2.3.2.2
Volume Densities of Mucin Granules

In animals fed on the commercial diet there was no significant difference in the volume density of mucin granules in either the small or large intestine. However, in conventional animals fed on a purified diet there was a significantly greater volume density of neutral mucin-containing cells of the small intestine.

3.2.3.2.3
Staining Density of Mucin Granules

In the small intestine of rats fed on the commercial diet the staining densities of neutral and acid mucin-containing granules were significantly increased in conventional rats when compared with those in germ-free rats (Fig. 7b). However, in rats fed on a purified diet the staining density of neutral mucin granules was significantly reduced in the cells of the small intestine but significantly increased in the cells of the large intestine when either the conventional or human flora was present.

In animals fed on the commercial diet the staining density of neutral and of acid mucin-containing cells was significantly increased in the small intestine and that of the sulphomucin-containing cells significantly reduced in the large intestine of the animals with a conventional flora when compared to that seen in the germ-free animals. When fed a purified diet the staining density of neutral mucin was significantly reduced in the large intestine in the presence of both the conventional and human flora. The presence of a human flora also significantly decreased the staining density of the sulphomucins in the small intestine and significantly increased the staining density in the large intestine.

3.2.3.3
Discussion

To extend the histochemical evidence described in Sect. 3.2.1 which found changes in the composition of intestinal mucins in response to two different diets and two different microbial floras, quantitative image analysis methods have been applied (Rubio et al. 1990; Phillips and Wilson 1993) to evaluate the amount and types of mucins present in the intestinal epithelial cells. As of this date, no morphometric studies have comprehensively assessed whether the mucin secretory activity of intestinal epithelial cells is influenced by the nature of the diet or by the presence or absence of a microbial flora in the intestinal lumen. The results of this morphometric investigation show that neutral, acidic and sulphomucins of the intestinal epithelial cells from the jejunum and from the proximal colon change in response to diet and/or microflora.

3.2.3.3.1
Effects of Diet

In the present study there was no difference in the number of mucin-containing cells in the germ-free animals as a result of differences in diet but when conventional rats were fed the purified diet, there was a reduction in the number of acidic carboxylated mucin-containing cells in the small intestine. However, when the diet is constant (i.e. purified), both conventional flora and human flora caused a reduction in the number of both acidic and sulphomucin-containing cells. Although it is difficult to know if it is the purified diet or the presence of flora that are responsible for the reduction in numbers, what is clear is that a combination of the two results in a more widespread reduction of mucin-producing cells in the mucosa.

Stereological morphometric studies by Lundin et al. (1993) indicate that although fibre supplementation to a low fibre diet did not effect the mucin quality of goblet cells, it did increase the number of goblet cells in the small intestine. Other studies have shown that indigestible dietary carbohydrates increased the number of all three types of mucin-containing goblet cells in the large intestine (Meslin et al. 1993). These different results may be due to the quantitative methods used and the site in the bowel from where the tissue was removed.

It is unclear as to the significance of the changes recorded in the present study of volume density and staining density of the mucin granules. Certainly in some situations, e.g. in the small intestine of germ-free rats fed a purified diet, it is possible to correlate an increased volume density with a reduced staining density suggesting a dilution of the mucus. However, this is not always the case. The increased staining density of neutral, acidic and sulphomucins in the large intestine of germ-free rats fed a commercial diet is not balanced by a reduction in volume density. It is possible therefore that the coarse nature of the fibre in the commercial diet may have improved the viscosity and storage of the mucin granules, which might explain the clinically observed benefits of high fibre diets (Kritchevsky 1988). Studies by Schneeman et al. (1982) and Vahouny et al. (1985) showed that a wheat bran-containing diet increased the mucin turnover and increased the incorporation of [3H]glucose and [35S]sulphate into intestinal glycoproteins thus indicating an increased differentiation of goblet cells. Histochemical studies of intestinal mucins in patients with inflammatory bowel disease have shown qualitative staining differences between sialomucins, sulphomucins and neutral mucosubstances (Filipe and Branfoot 1974; Boland et al. 1982; Ehsanullah et al. 1982; Jacobs and Huber 1985).

3.2.3.3.2
Effects of Flora

The presence of either flora reduced the number of mucin-containing cells in the small intestine of rats fed on the purified diet. Although the volume density of the neutral mucin-containing cells was increased, this was compensated for by a reduction in their staining density. However, in the large intestine the staining density of neutral mucin-containing cells was increased.

The changes in the mucin-containing cells of HFA rats found by morphometric methods when correlated with quantitative morphological data described in Sect. 3.2.2 suggest that the changes in the staining density of neutral and sulphomucins is associated with alterations in the villus and crypt lengths of the small intestine and the crypt lengths of the large intestine. Changes in the number, volume and staining densities of mucin-containing cells may not only be influenced by mucin-degrading properties of bacterial glycosidases but also by indirect, secondary effects of the microbial flora on the differentiation of mucin-secreting cells from the crypt epithelia. Although it is known that little bacterial degradation of mucus glycoproteins occurs in the small intestine under normal conditions due to the sparse populations of indigenous bacteria (Hoskins 1978), the mechanism of differentiation of goblet cell from the crypt stem cell in response to the microbial inoculation is still not understood. The results of this study support the suggestion that differences in the intestinal mucins of conventional and HFA rats might be due to interactions between components of the diet and the microflora (Ducluzeau 1984). Although it cannot be claimed that all components of the indigenous human gut flora were present in the HFA rats, the differences observed in mucin composition between HFA and conventional rats cast doubt on the suitability of conventional animals as a model for man in this type of study. Further exploration of the use of rats harbouring floras of human origin seems warranted.

In conclusion, this study has shown that image analysis offers a sensitive method for assessing the quantitative changes in mucin secretory activity of intestinal epithelial cells with different dietary patterns in the presence of an indigenous or a human microbial flora. In the jejunum the fibre-rich commercial diet was particularly influential in modifying the acidic mucins in the presence of an indigenous flora. This suggests that the quantity and composition of mucus produced by the intestinal epithelial cells are more influenced by diet in the small intestine and by microflora in the large intestine. More in-depth studies are needed to reveal the effects of interactions between diet and microflora on the goblet cell mucins. Further studies of goblet cells described in Sect. 3.2.4 using lectins as tools may indicate whether alterations in glycoconjugate expression accompany changes in composition of intestinal mucins.

3.2.4
Lectin-Binding Pattern of Mucins

The epithelial structure of the mammalian intestinal mucosa reflects the absorptive and secretory activity of its various segments. The apical surface membrane of intestinal epithelial cells possesses a well-defined layer of glycoproteins and glycolipids, the glycocalyx, which is involved in key cellular processes such as adhesion, antigenic recognition, differentiation, transport and malignant transformation (Ashwell and Morell 1977; Strous and Dekker 1992). In addition to this glycocalyx, the luminal surface of the gastrointestinal tract is also covered with mucin-type glycoproteins secreted by the goblet cells. Since the capacity of mucin-type glycoproteins to form a protective gel (Allen and Pearson 1993) and to interact with bacterial species depends largely on their content of oligosaccharides (Mantle and Husar 1994), it is generally accepted that changes in the secretory glycoconjugates result in impairment of these functions and are directly related to some intestinal diseases (Jacobs and Huber 1985). It is also known that mucins have an aggregating effect on several bacterial species (Piotrowski et al. 1994) and that the carbohydrate moieties of intestinal mucin prevent attachment of pathogenic bacteria by modulating their adherence to the intestinal epithelium (Ravidin et al. 1985; Mantle and Husar 1994). It would seem pertinent, therefore, to investigate whether or not intestinal luminal factors by their contact with the epithelial layer may modify the mucin composition and thus protect the gastrointestinal tract from various diseases. Studies described in Sects. 3.2.1, 3.2.2 and 3.2.3 have shown that dietary constituents and microflora alter the crypt-villus architecture and modify the mucin composition of the intestinal tract. In order to understand the altered status of mucin composition in response to dietary changes and to evaluate the relation between the mucin and microflora, a more detailed analysis of the effects of diet and/or microflora on the carbohydrate sequences of the intestinal mucins is required.

Lectins, glycoproteins of plant or animal origin, have the ability to bind specific carbohydrate residues of cell glycoconjugates, particularly in terminal positions (Damjanov 1987). Previous studies have used fluorescein-conjugated lectins as specific molecular probes to localise differences in the distribution of secretory glycoconjugates and to detect the maturational changes in glycosylation of mucin during the differentiation of goblet cells from the base of the crypt to the mucosal surface in the rat small intestine (Etzler and Branstrator 1974) and colon (Freeman et al. 1980; Brasitus et al. 1982). Lectin histochemistry has also been pertinent in studying the alteration in secretory glycoconjugates of goblet cells in animal models of intestinal diseases (Boland and Ahnen 1985; Itagaki et al. 1994) and during the pathogenesis of human colorectal diseases such as carcinoma (Egea et al. 1993) and inflammatory bowel disease (Jacobs and Huber 1985) and is therefore well suited to analyse the gastrointestinal carbohydrate changes in situ.

The aim of the experiments conducted for this study was to examine the effects of the interactions between the dietary constituents and/or microflora on goblet cell glycoconjugates along the intestinal tract. To determine the specific changes in goblet cell glycoconjugates, their terminal carbohydrate residues were determined using lectins with carbohydrate specificities for five different sugars. Segments from the jejunum and the colon of the rat under different dietary regimens/microbial floras were investigated.

3.2.4.1
Controls

In all sections incubated in the presence of specific inhibitory monosaccharides the lectin binding was inhibited or greatly diminished after this absorption. Con A labelling of goblet cell mucins was not observed in any animal and therefore will not be mentioned again in the Results section.

3.2.4.2
Effects of Diet

The influence of the commercial or the purified diet on the lectin binding of goblet cell mucins in GF rats is shown in Table 7 and of the CV rats in Table 8.

Table 7. Lectin reactivity of goblet cell mucins in germ-free rats given a commercial rodent diet (*GF-C*) or a diet of purified ingredients (*GF-P*)

Intestine and	GF-C					GF-P				
cell type	Con A	WGA	DBA	MAA	UEA- I	Con A	WGA	DBA	MAA	UEA- I
Small intestine										
Surface goblet cells	0	+++	+[a]	+++	+++	0	++	0[a]	++	+
Upper crypt goblet cells	0	+++	++	++	++	0	++	0	++	++
Lower crypt goblet cells	0	++	0	+++	++	0	++	0	+++	++
Large intestine										
Surface goblet cells	0	++	+	++	+	0	++	0	+	++
Upper crypt goblet cells	0	+++	++[a]	++	++	0	++	0[a]	++	+
Lower crypt goblet cells	0	+++	+[a]	+++	(+)	0	+++	(+)[a]	+++	(+)

Intensity of reaction: 0 no reactivity; (+) weak reactivity; + moderate reactivity; ++ strong reactivity; +++ intense reactivity (for abbreviations of lectins see Table 2 and of experimental groups see Materials and Methods).
[a] Lectin binding enhanced by neuraminidase pretreatment.

3.2.4.2.1 Small Intestine

The labelling with WGA and MAA was more intense in the surface goblet mucins of groups fed the commercial diet. In the GF animals DBA binding was present in the surface goblet cells of animals fed a commercial diet (Fig. 9a) but not in those fed a purified diet whereas in the CV animals the DBA lectin binding was reversed to that seen in the GF animals. Following neuraminidase treatment, however, DBA binding was seen in the surface goblet cells of the GF and CV animals which were previously negative. In addition, DBA binding was more enhanced in those goblet cells which were previously positive. In the GF animals fed a commercial diet the surface goblet cells were more intensely stained with UEA-I when compared to that seen in the surface goblet cells of other groups.

3.2.4.2.2 Large Intestine

Although differences between the dietary groups were observed for MAA and UEA-I binding, the most obvious difference was seen in the DBA binding of both surface and crypt goblet cells which was much greater in animals fed the commercial diet in both the GF (Fig. 9b,c) and CV animals.

Table 8. Lectin reactivity of goblet cell mucins in conventional rats or rats associated with a human flora

Intestine and cell type	CV-C					CV-P					HFA-P				
	ConA	WGA	DBA	MAA	UEAI	ConA	WGA	DBA	MAA	UEAI	ConA	WGA	DBA	MAA	UBAI
Small intestine															
Surface goblet cells	0	+++	0[a]	++	+	0	+	+[a]	(+)	+	0	++	0[a]	++	++
Upper crypt goblet cells	0	++	0[a]	+++	++	0	++	(+)[a]	0	++	0	++	0	++	++
Lower crypt goblet cells	0	++	0	++	++	0	++	0	0	+++	0	++	0	+++	+
Large intestine															
Surface goblet cells	0	++	(+)	++	+	0	+	0	(+)	++	0	++	0	+	0
Upper crypt goblet cells	0	++	++	++	+	0	++	0	+	+	0	++	+	++	+
Lower crypt goblet cells	0	+++	+	+++	++	0	+++	(+)	++	(+)	0	+++	+	+++	+++

Intensity of reaction: 0 no reactivity; (+) weak reactivity; + moderate reactivity; ++ strong reactivity; +++ intense reactivity (for abbreviations of lectins see Table 2 and of experimental groups, see Materials and Methods).

[a] Lectin binding enhanced by neuraminidase treatment

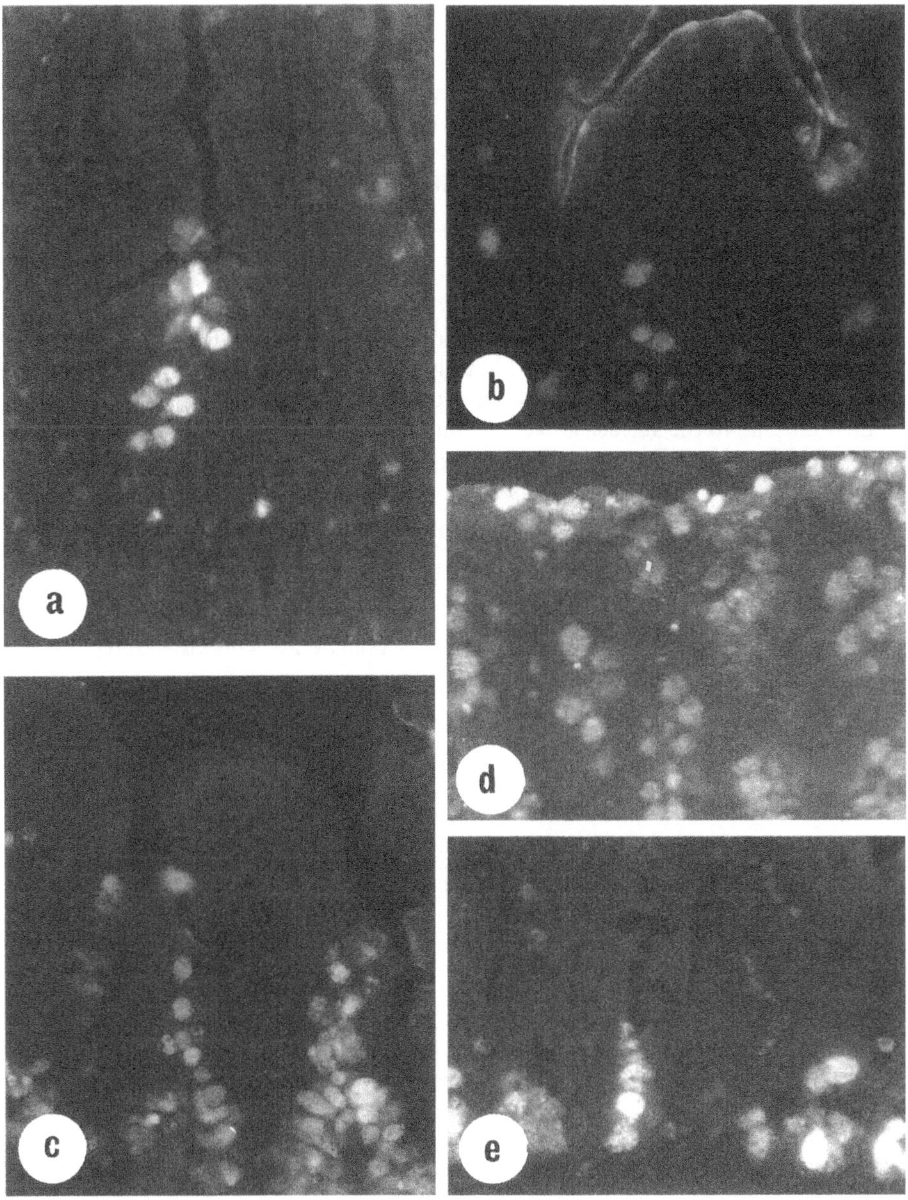

Fig. 9a–e. Binding of lectins to goblet cell mucins. **a** FITC-DBA labelling of small intestine in a germ-free rat fed the commercial diet. Goblet cells in the surface epithelium and in the upper crypt show a *bright fluorescence.* ×330. **b** FITC-DBA labelling in the large intestine of a germ-free rat fed the purified diet. The goblet cell mucin in the lower part of the crypt is labelled whereas those in the upper part and the surface epithelium are almost negative. ×265. **c** Unlike the GF-P rat, mucin granules in the goblet cells of germ-free rats fed the commercial diet are positive in the upper and lower portions of the crypt. ×265. **d** FITC-UEA-I binding in a germ-free rat fed the purified diet. Surface goblet cells show a strong affinity for UEA-I. **e** Binding of UEA-I to the large intestine in a rat inoculated with a human flora. The goblet cells mucin in the crypt is labelled, whereas that in the surface epithelium is negative. ×265

3.2.4.3
Effects of Conventional Microflora

The effects of conventional microflora and of the interactions of the diet and indigenous flora on the expression of various carbohydrate residues were examined by comparing the lectin-binding pattern of GF-C and CV-C rats and of GF-P and CV-P rats (Tables 7, 8).

3.2.4.3.1
Small Intestine

The WGA or MMA lectin binding to goblet cell mucins of GF and CV rats fed a commercial diet were similar. DBA binding of the surface and crypt goblet cells was present in the GF animals but only after neuraminidase treatment in the CV animals. UEA-I binding to the surface goblet cell mucins of the CV animals was also reduced when compared to that seen in the GF animals.

When fed a purified diet animals with a conventional flora revealed considerably less MAA binding to goblet cells

3.2.4.3.2
Large Intestine

No major differences in lectin binding were observed between the GF-C and CV-C rats and GF-P and CV-P animals.

3.2.4.4
Effects of Human Microflora

The effects of human flora were examined by comparing the lectin binding pattern between the GF-P and HFA-P rats (Tables 7, 8).

3.2.4.4.1
Small Intestine

Apart from minor changes in the UEA-I binding to goblet cell mucins there were no other differences observed in the small intestine.

3.2.4.4.2
Large Intestine

As in the small intestine there were differences in the binding of UEA-I. The surface goblet cells showed no UEA-I binding sites in the presence of human flora but in GF animals UEA-I binding intensity was strong (Fig. 9d). However, in the deeper crypt cells the UEA-I binding was considerably increased in animals containing a human flora (Fig. 9e).

3.2.4.5
Discussion

Lectins were used as histochemical probes to look for changes in the expression of carbohydrate residues in goblet cell mucins in response to diets and microflora. The strong binding of WGA and UEA-I to the goblet cells in both small and large intestine indicates that mucins are rich in *N*-acetyl-glucosamine and fucose, both of which are typical carbohydrate residues for mucin-type glycoproteins (Strous and Dekker 1992). The absence of Con A binding in all experimental groups is in agreement with the finding that the goblet cells do not synthesise large amounts of mucins with mannose and glucose sugar residues (Forstner et al. 1982). Differentiation-related changes in the carbohydrate residues of mucins were evident from the differences in the intensity of lectin binding to goblet cells in the upper and lower crypts and in the surface epithelium. Similar alterations in secretory glycoconjugates during the maturation of goblet cells from base of the crypt to the luminal surface have been reported in rat (Etzler and Branstrator 1974; Freeman et al. 1980a,b; Brasitus et al. 1982), pig (Moré et al. 1980), monkey (Posel et al. 1988) and human (Filipe and Fenger 1979; Boland et al. 1982; Gorelick et al. 1982; Jacobs et al. 1983) intestine and are therefore not an unexpected finding. However, a new and notable finding is that the UEA-I, MAA and DBA binding for Fuc, sialic acid and GalNAc residues, respectively, are influenced by the nature of the diet and the presence of the gut microflora.

3.2.4.5.1
Effects of Diet

The increased staining intensity of WGA and MAA particularly in the surface goblet cells of GF and CV rats fed on the commercial diet would indicate that the commercial diet induces an addition of terminal GlcNac and sialic acid linked d-galactose residues to the glycoconjugate core of intracellular mucins. These differences in the lectin binding pattern are probably due to the coarse nature of the fibre in the commercial diet which induced the expression of various glycosyltransferases responsible for the transfer of sugar residues to the side chains of mucins. Likewise, the fibrous nature of the commercial diet probably enhances the protection of the mucosa by increasing the viscosity of the mucins (Strous and Dekker 1992) and by increasing the resistance of glycoproteins to proteases (Schauer 1985), which would be explained by the addition of further carbohydrate residues. The differences observed in the DBA binding sites as a result of different diets and flora are difficult to explain. When the sialic acid residues were removed by neuraminidase digestion, numerous DBA binding sites were exposed in the goblet cells of the small intestine of both germ-free and conventional rats but only in the large intestine of germ-free rats. The likelihood is that neuraminidase released from the microflora (Allen and Starkey 1974) is normally present in the colonic mucosa of conventional animals. However, the more marked DBA binding in the colonic mucosa of both germ-free and conventional rats fed the commercial diet could indicate that differences in diet can modify the GalNAc residues on colonic goblet cell mucin as well.

A number of mechanisms by which the diet-microbial interactions may influence the pattern of glycosylation of mucins and consequently the lectin binding should be considered: (1) an increase in the short chain fatty acid (SCFA) production from the fermentation of dietary fibre which then lowers the luminal pH (Meslin et al. 1993) in the large intestine of CV-C animals and increases the amount of glycoconjugates with terminal GalNAc in comparison to CV-P animals; (2) stimulation of the crypt cell production rate (Sakata 1987) in the colon of CV-C animals, which alters the differentiation of goblet cells and increases the amount of GlcNAc and sialic acid residues.

3.2.4.5.2
Effects of Flora

The presence of conventional microflora reduced the MAA binding sites along the intestinal crypts in rats fed on the purified diet. In addition, the human flora considerably altered the UEA-I binding in the same location under the same diet. The mechanisms involved in an increase in UEA-I binding sites from the crypt to surface in goblet cells of the small intestine and a decrease in the large intestine of rats containing a human flora are not yet clear. The differences in MAA and UEA-I binding might be accounted for by different characteristics of gut flora and further studies will be necessary to evaluate the sialidase and fucosidase-producing properties of the conventional and human microflora. It is possible that changes in expression of carbohydrate residues of gut mucins are modified by interactions between lectin-like substances on the membrane of microorganisms and glycoconjugates located on the epithelial surface (Ofek and Sharon 1988). Analysis of glycosidase profile of microbial flora (Hoskins et al. 1985) and investigations into the epithelial attachment of gut flora may clarify the precise interpretation of the lectin histochemical data in the present study and may help in the understanding of microbial–host interactions in health and disease.

The conclusions arising from this study are that the interactions between diet and microflora on the epithelial surface of the intestinal mucosa modify the glycoconjugate composition of goblet cells and that this modification may be attributed to the changes in the activity of several glycosyltransferases and glycosidases as also suggested by Freeman et al. (1980). However, displaying of specific carbohydrate residues in incompletely glycosylated mucins may also be responsible for similar changes. These differences in the carbohydrate residues of goblet cell secretion in this experimental model have provided valuable information on the influence of diet on mucin-type glycoproteins. This is vital for the understanding of glycosylation changes of the mucosa in various intestinal diseases (Boland et al. 1982; Jacobs and Huber 1985).

4 Influence of Diet and Microflora on Distribution of Enteroendocrine Cells

Enteroendocrine cells are scattered in the epithelium of the mammalian gastrointestinal tract, and each individual cell type is capable of producing and secreting a specific biologically active peptide and/or amine (for reviews, see Solcia et al. 1987; Polak 1989). Studies using immunohistochemical methods have identified at least 16 different subpopulations of enteroendocrine cells based on the nature of their stored peptides and biogenic amines and each maintains a distinctive distribution along the gastrointestinal tract from stomach to the anal canal (for review, see Sundler et al. 1993). Since most of the enteroendocrine cells in the intestinal tract are of 'open type' (Krause et al. 1985) and reach the gut lumen by apical processes, it has been suggested that these cells may respond to intraluminal stimuli such as pH and nutrients (Sundler et al. 1993). A few studies are available in which the in vivo release of gastroenteric peptides have been shown to be influenced by the ingestion of proteins and lipids (Ensinck et al. 1990; Barber et al. 1991) and by the presence of the conventional flora in the intestinal lumen (Uribe et al. 1994). Investigations using in vivo techniques, however, must be correlated with the analysis of changes in gastroenteric peptides in situ.

Much of the research in this review, on the effects of luminal stimuli on intestinal mucosa, has been concerned with the modification of the diet and bacterial flora. The results described in Sect. 3 provide evidence that the diets containing fibre and the presence of flora in the intestinal lumen modify the intestinal morphology and alter the composition of intestinal mucins. However, the precise mechanism by which these changes occur in the gastrointestinal mucosa is not known and whether the activity of enteroendocrine cells is influenced by the interactions between the diet and gut microflora in the intestinal lumen remains to be investigated.

In the past, investigating the in situ changes in endocrine cell population in response to luminal stimuli has been limited by a lack of an appropriate model system. The use of germ-free rats fed different diets and harbouring either a conventional flora or a flora of human origin has provided the means for characterising the changes in the synthesis and secretion of intestinal mucins. The investigations described in this study utilise the rat experimental model to evaluate the effects of diet and microflora on distribution of enteroendocrine cells along the gastrointestinal tract. Since enteroendocrine cells contain biogenic amines in addition to their peptides, this investigation was focused on quantification of these cells by using the Grimelius argyrophilic method and immunohistochemistry using an antiserum against serotonin. The Grimelius technique stains most endocrine cell types except the peptide YY cells and somatostatin and cholecystokinin producing cells (Dayal 1991). Serotonin is not localised in all enteroendocrine cells and is present in non-peptide producing endocrine cells (Cetin 1990). By the combination of these two methods, the vast majority of enteroendocrine cells should be detected.

4.1 Results

The number of argyrophil endocrine cells in the surface epithelium, upper crypt and lower crypt of the jejunum and colon are shown in Figs. 10 and 11 and the number of serotonin immunoreactive (SIR) cells in Figs. 12 and 13. Figure 14 shows the total number of argyrophil and SIR cells in the jejunum and Fig. 15 in the colon. The total number denotes the sum of surface and crypt cells in the unit length of the mucosa.

4.1.1 Effects of Diet

The influence of diet was assessed by comparing the germ-free and conventional rats receiving either the commercial or the purified diet.

4.1.1.1 Argyrophil Endocrine Cells

4.1.1.1.1 Jejunum

The number of argyrophil endocrine cells in the surface epithelium of the jejunum was significantly less in the germ-free rats fed on the commercial diet when compared to those fed on the purified diet (see Fig. 10a). In conventional rats, the number was significantly less in the lower crypt of the jejunum in animals fed the commercial diet as compared to those fed on the purified diet (see Fig. 10b). There were numerous argyrophil endocrine cells in the lower and upper crypt as well as surface epithelium of jejunum in the conventional rats fed the purified diet (Fig. 16a). The highest total number of argyrophil endocrine cells was found in germ-free animals fed the purified diet (Fig. 14).

4.1.1.1.2 Colon

The numbers of argyrophil endocrine cells in the colon of germ-free and conventional rats fed on two different diets were not significantly different from one another (see Fig. 10a,b).

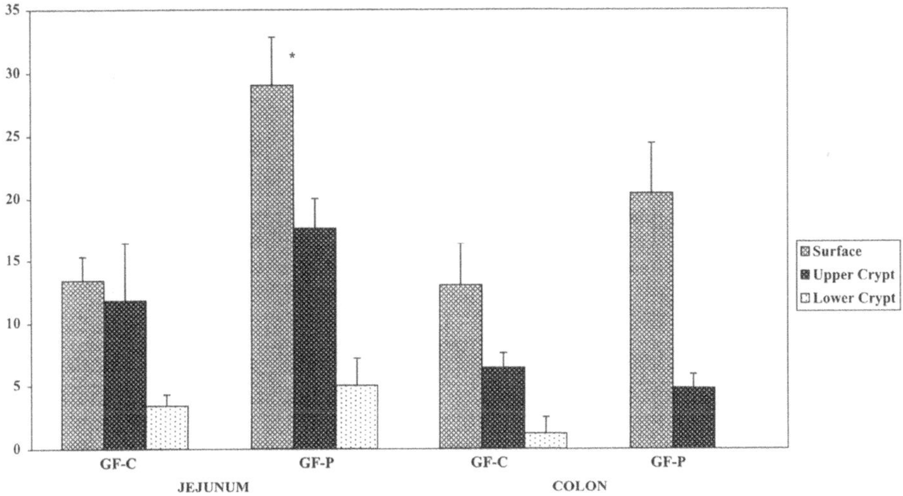

a * Significantly different at P<0.01 from corresponding value from germ-free rats fed a commercial diet

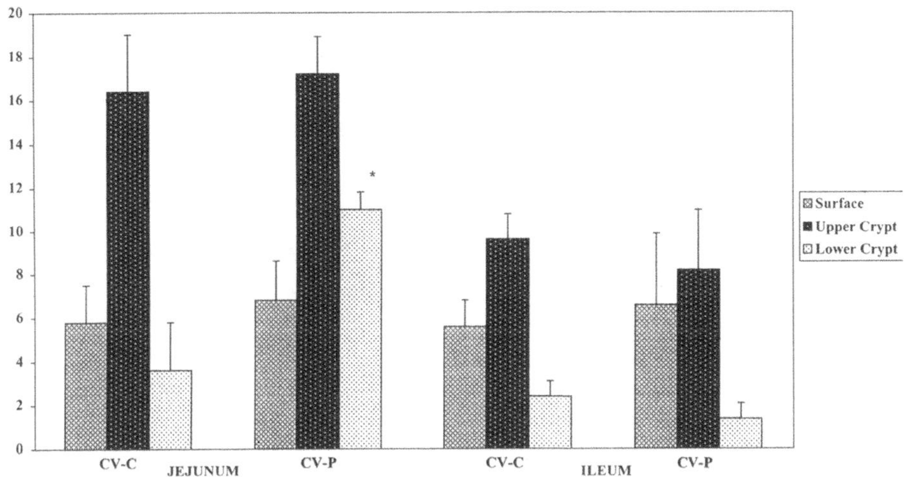

b * Significantly different at P<0.02 from corresponding value for conventional rats fed a commercial diet

Fig. 10. a Effects of diet on number of argyrophil endocrine cells in germ-free rats (values are means±SEM). *Significantly different at $P<0.01$ from corresponding value from germ-free rats fed a commercial diet. **b** Effects of diet on number of argyrophil endocrine cells in conventional rats (values are means±SEM) *Significantly different at $P<0.02$ from corresponding value for conventional rats fed a commercial diet

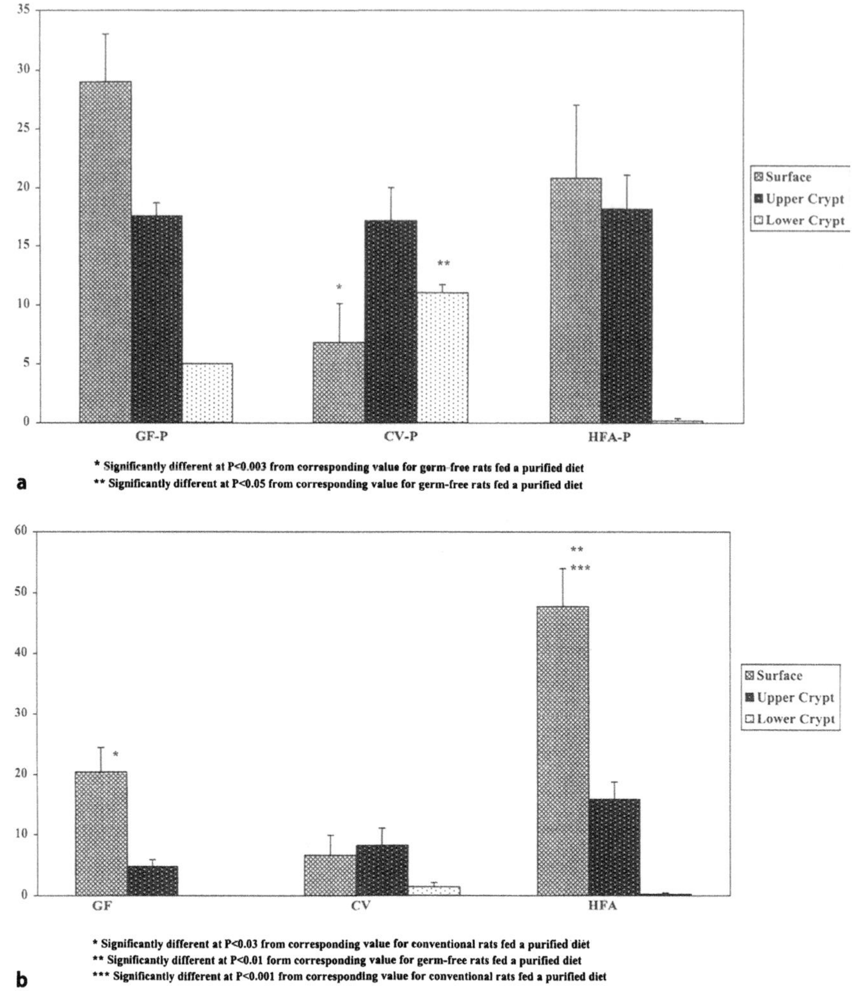

Fig. 11. a Effects of microflora on number of argyrophil endocrine cells in the jejunum of rats fed the purified diet (values are means±SEM). *Significantly different at $P<0.003$ from corresponding value for germ-free rats fed a purified diet. **Significantly different at $P<0.05$ from corresponding value for germ-free rats fed a purified diet. **b** Effects of microflora on number of argyrophil endocrine cells in the colon of rats fed the purified diet (values are means±SEM). *Significantly different at $P<0.03$ from corresponding value for conventional rats fed a purified diet. **Significantly different at $P<0.01$ from corresponding value for germ-free rats fed a purified diet. ***Significantly different at $P<0.001$ from corresponding value for conventional rats fed a purified diet

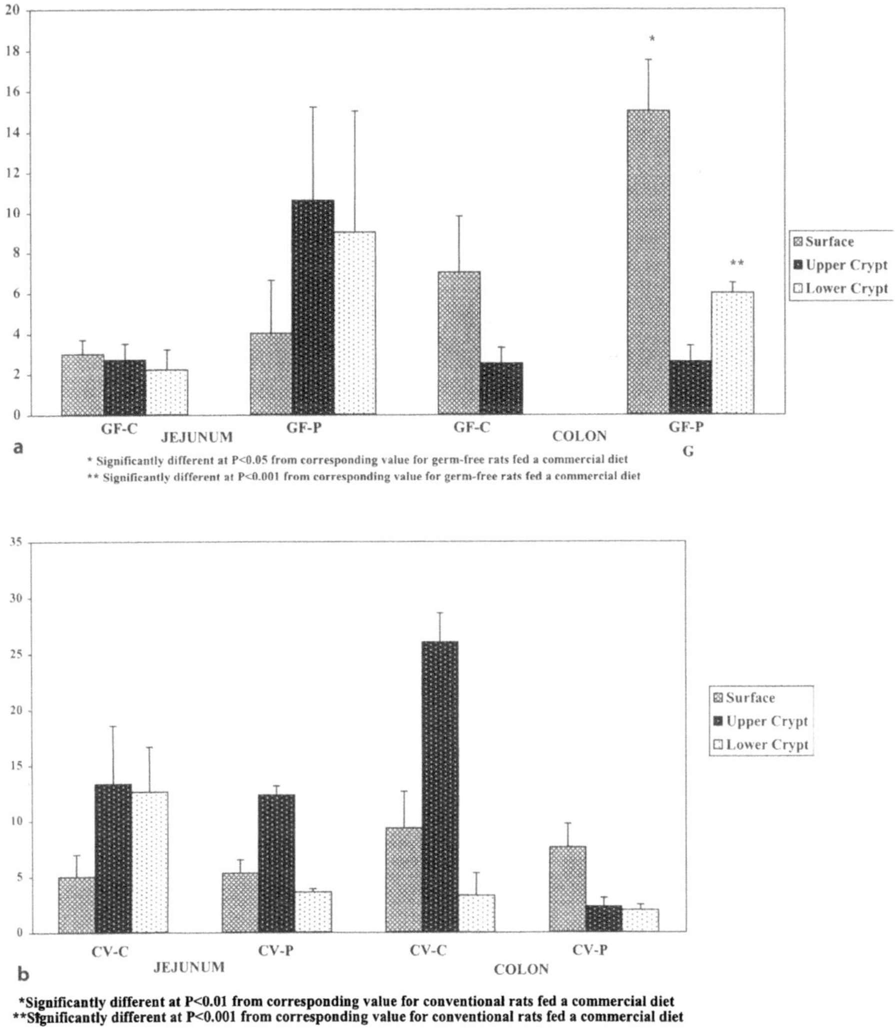

Fig. 12. a Effects of diet on number of serotonin immunoreactive cells in conventional rats (values are means±SEM). *Significantly different at $P<0.05$ from corresponding value for germ-free rats fed a commercial diet. **Significantly different at $P<0.001$ from corresponding value for germ-free rats fed a commercial diet. **b** Effects of diet on number of serotonin immunoreactive cells in conventional rats (values are means±SEM). *Significantly different at $P<0.01$ from corresponding value for conventional rats fed a commercial diet. **Significantly different at $P<0.001$ from corresponding value for conventional rats fed a commercial diet

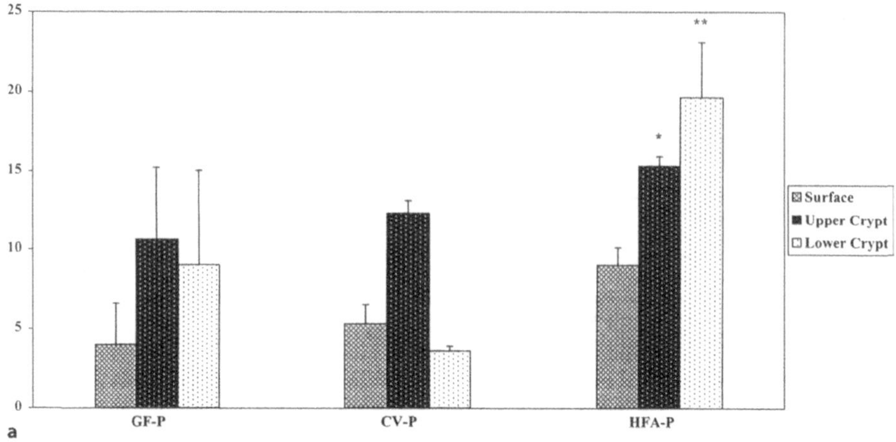

a

* Significantly different at P<0.05 from corresponding value for conventional rats fed a purified diet

** Significantly different at P<0.01 from corresponding value for conventional rats fed a purified diet

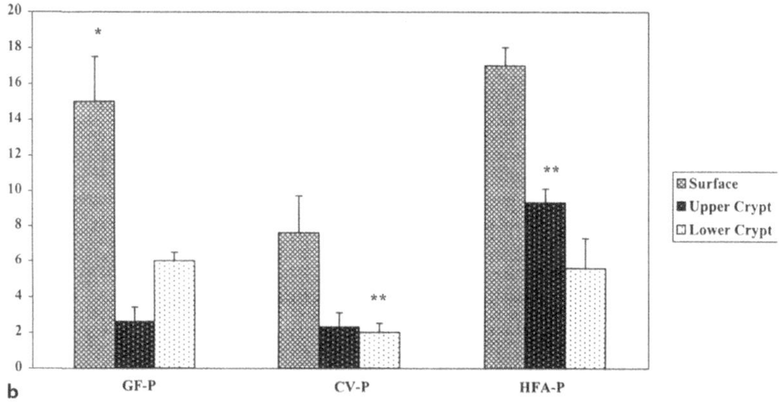

b

*Significantly different at P<0.05 from corresponding value for germ-free rats fed a commercial diet

**Significantly different at P<0.005 from corresponding value for germ-free rats fed a purified diet

Fig. 13. a Effects of microflora on number of serotonin immunoreactive cells in the jejunum of rats fed the purified diet (values are means±SEM). *Significantly different at $P<0.05$ from corresponding value for conventional rats fed a purified diet. **Significantly different at $P<0.01$ from corresponding value for conventional rats fed a purified diet. **b** Effects of microflora on number of serotonin immunoreactive cells in the colon of rats fed the purified diet (values are means±SEM). *Significantly different at $P<0.05$ from corresponding value for germ-free rats fed a commercial diet. **Significantly different at $P<0.005$ from corresponding value for germ-free rats fed a purified diet

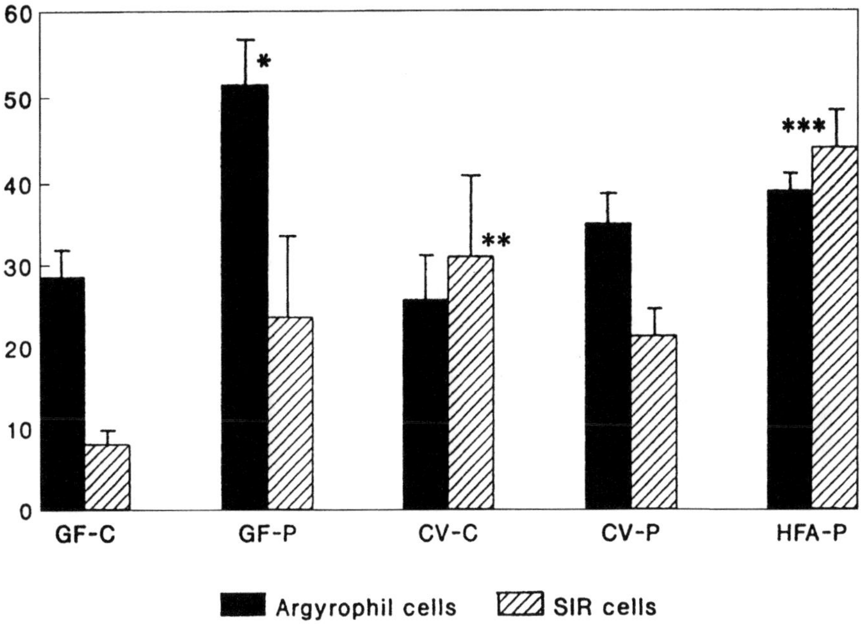

Fig. 14. Effects of diet and microflora on the total number of enteroendocrine cells in the jejunum. Values are expressed as means±SEM per experimental group. For abbreviations of experimental groups see Materials and Methods. *Significantly different at $P<0.02$ from the corresponding value for germ-free rats fed the commercial diet. **Significantly different at $P<0.05$ from corresponding value for germ-free rats fed the commercial diet. ***Significantly different at $P<0.01$ from the corresponding value for conventional rats fed the purified diet

4.1.1.2
SIR Cells

4.1.1.2.1
Jejunum

Apart from a significantly higher number of SIR cells in the lower crypt in conventional rats fed the commercial diet no significant differences in the number of SIR cells were observed in the jejunum between the two germ-free and the two conventional dietary groups (see Fig. 12a,b).

4.1.1.2.2
Colon

The number of SIR cells was significantly higher in the surface epithelium and lower crypt of germ-free rats fed the purified diet compared to those fed the commercial diet (see Fig. 12a). However, in conventional animals, the number of cells in the upper crypt was significantly higher in the group fed the commercial diet than in their counterparts fed on the purified diet (see Fig. 12b). In germ-free rats fed the purified diet, the total number of SIR cells was significantly higher than in their GF-C counter-

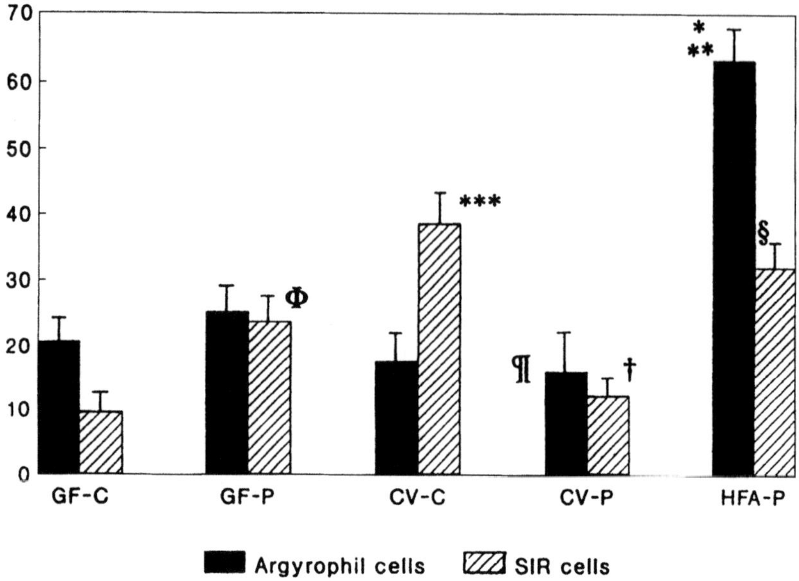

Fig. 15. Effects of diet and microflora on the total number of enteroendocrine cells in the colon. Values are expressed as means±SEM per experimental group. For abbreviations of experimental groups see Materials and Methods. *Significantly different at $P<0.005$ from the corresponding value for germ-free rats fed the purified diet. **Significantly different at $P<0.001$ from the corresponding value for conventional rats fed the purified diet. ***Significantly different at $P<0.001$ from the corresponding value for germ-free rats fed the commercial diet. ΦSignificantly different at $P<0.01$ from the corresponding value for germ-free rats fed the commercial diet, ¶at $P<0.05$ from the corresponding value for conventional rats fed the commercial diet, †at $P<0.05$ from the corresponding value for germ-free rats fed the purified diet, §at $P<0.01$ from the corresponding value for conventional rats fed the purified diet

parts. However, the number of SIR cells was significantly lower in conventional animals fed the purified diet than those fed the commercial diet (see Fig. 15).

4.1.2 Effects of the Microflora and Diet–Microflora Interactions

The effects of conventional flora and the effects of the interactions between the diet and the indigenous flora on the number of enteroendocrine cells were assessed by comparing the data from germ-free and conventional rats receiving the commercial diet and of germ-free and conventional rats receiving the purified diet. The effects of human flora were assessed by comparing germ-free and HFA rats receiving the purified diet. In groups fed on the purified diet, the effects of the two different microfloras were assessed by comparing the data between the conventional rats and rats associated with human flora.

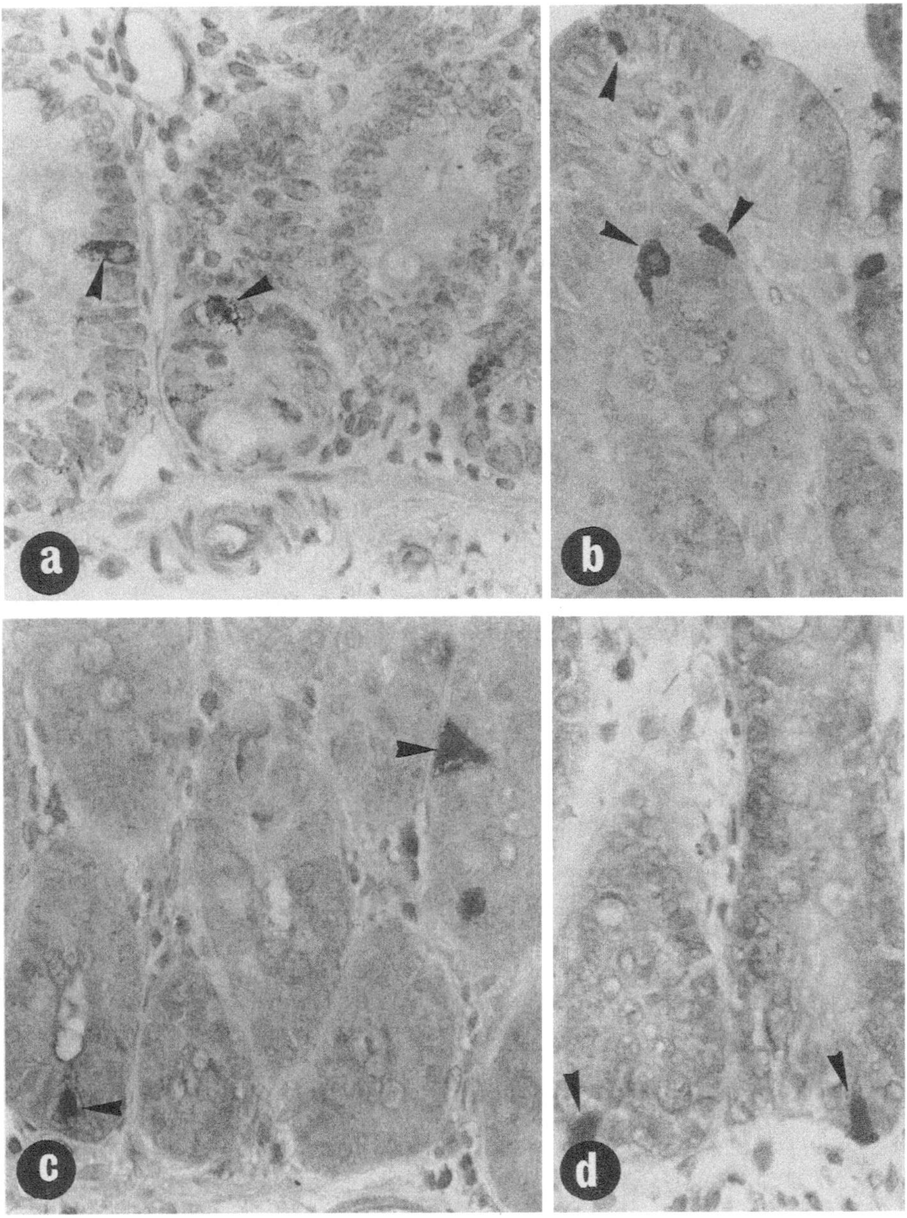

Fig. 16a–d. Enteroendocrine cells in the rat intestine. **a** Argyrophil endocrine cells in the jejunum of a conventional rat fed the purified diet. Argyrophil cells (*arrows*) are seen in the crypt epithelium. ×330. **b** Colonic crypts of an HFA rat fed the purified diet showing argyrophil endocrine cells (*arrows*) in the surface epithelium and upper crypts. ×265. **c** Jejunum of a conventional rat fed the commercial diet. Serotonin immunoreactive cells (*arrows*) are seen in the lower crypts. ×530. **d** Colonic crypts of a germ-free rat fed the purified diet showing serotonin immunoreactive cells (*arrows*) in the lower crypt. ×530

4.1.2.1
Argyrophil Endocrine Cells

4.1.2.1.1
Jejunum

When the germ-free and conventional rats were compared, there was a reduced number of surface argyrophil endocrine cells in the CV rats fed both commercial ($P<0.01$) and purified diets (see Fig. 11a). There was a slightly higher number of endocrine cells in the lower crypts of animals with a conventional flora. No significant differences in the number of argyrophil endocrine cells were found between the germ-free and HFA animals.

4.1.2.1.2
Colon

There were no significant differences observed between the number of argyrophil endocrine cells in the colonic mucosa when germ-free and conventional rats fed the commercial diet were compared. However, the number of surface argyrophil cells (Fig. 16b) in the germ-free and HFA rats were significantly higher then the number seen in the conventional rats (see Fig. 11b). In HFA rats, the total number of argyrophil endocrine cells in the colon was significantly higher when compared with germ-free and conventional rats fed the purified diet (Fig. 15).

4.1.2.2
SIR Cells

4.1.2.2.1
Jejunum

When germ-free and conventional rats fed the commercial diet are compared for the effect of flora on number of SIR cells, the only significant difference observed was in the higher number of lower crypt cells in the CV rats ($P<0.03$; Fig. 16c). The total number of SIR cells was also significantly higher in the conventional rats when compared with the germ-free rats fed the commercial diet (Fig. 14). When the effects of flora on rats fed the purified diet were compared, there was a significantly higher number of SIR cells in the lower parts of crypt in the presence of human flora (see Fig. 13a). The total number of SIR cells was significantly higher in HFA rats compared with their conventional counterparts (Fig. 14).

4.1.2.2.2
Colon

In the colon the effects of flora were less obvious. When germ-free and conventional rats fed a commercial diet were compared, there was a significantly higher number of SIR cells in the upper crypts in the presence of conventional flora ($P<0.002$). The total number of SIR cells was also higher in conventional animals fed the commercial diet (Fig. 15). In rats fed the purified diet the conventional flora reduced the number of SIR cells in the lower crypt (Fig. 16d) and the human flora increased in the upper crypt. The presence of conventional flora in animals fed the purified diet reduced the total number of SIR cells (Fig. 15).

4.2 Discussion

The enteroendocrine cells provide a source of polypeptide hormones and biogenic amines which are involved in modulation of secretory, absorptive and motor functions of the gastrointestinal tract (Dayal 1991). Evidence that enteroendocrine cells respond to changes in the luminal environment has emerged from in vitro experiments with nutrients (Barber 1993). Similarly, immunohistochemical studies involving transposition of intestinal segments have shown the effects of luminal signals on the expression of enteroendocrine products along the intestinal tract (Aiken et al. 1994). The results of in situ analysis of the occurrence and distribution of enteroendocrine cells in different parts of the small and large intestine presented in this study indicate that the number of enteroendocrine cells of the jejunum and proximal colon is also influenced by the dietary composition, presence or absence of microflora and the interactions between the diet and microflora in the intestinal lumen.

The regional differences in the distribution of enteroendocrine cells are well documented (Solcia et al. 1987) and the results of the present study confirm previous observations that enteroendocrine cells are more numerous in the small intestine than in the large intestine (Sjölund et al. 1983; Roth et al. 1990). In addition, the present results also demonstrate that diet–microbial interactions alter the number and distribution pattern of argyrophil cells and the number and distribution pattern of serotonin immunoreactive cells along the length of the intestinal tract and along the crypt-surface axis. For instance, in germ-free animals fed the commercial diet there is a decrease in the number of surface and total argyrophil endocrine cells in the jejunum and of serotonin immunoreactive cells in the colon. However, the commercial diet fed to conventional animals results in an increase in the upper crypt and the total number of serotonin immunoreactive cells in the colon. These findings indicate that the interactions between dietary components and microflora regulate the peptide synthesis of enteroendocrine cells in the small intestine and of biogenic amines in the large intestine. In the quantitative morphological study described in Sect. 3 there was a marked increase in the villus and crypt lengths in the jejunum and a decrease in the length of colonic crypts of germ-free and conventional rats fed the commercial diet. Although the precise regulatory mechanisms by which dietary components modify the cell proliferation of intestinal mucosa are not known, the results presented in this chapter, when correlated with the previous quantitative morphological data, suggest that the mucosal response to diet may be mediated by peptides and biogenic amines synthesised by the enteroendocrine cells. These observations are consistent with the findings that in rats fed different diets the gastrointestinal peptides such as gastrin, PYY and enteroglucagon play a role in altering the crypt cell proliferation rates (Goodlad et al. 1987; Johnson et al. 1988).

A recent study comparing germ-free and conventional animals fed the same diet has shown an increased number of serotonin-containing cells in the distal small intestine and colon of rats harbouring a rat flora (Uribe et al. 1994). The results of the present study have shown that in conventional animals the number of serotonin immunoreactive cells is higher in the jejunum and colon of animals fed the commercial diet and lower in those fed the purified diet. The observations in the present study indicate that the effects of microflora on the distribution of the enteroendocrine cells depend on the luminal interactions between the flora and dietary constituents and not on the flora alone as recently described (Uribe et al. 1994). Although the significance of these effects is not yet clear, the possibility exists that diet and flora may influence intestinal motility by modifying the gastroenteric products. The microflora (Huseby et al. 1990) and the release of somatostatin (Efendic and Mattsson 1978) and motilin (Strunz et al. 1976) have been implicated as luminal and systemic factors involved in gastroenteric motility. Although the effect of conventional flora on the number of enteroendocrine cells was dependent on the diet, the differences between the conventional and HFA rats fed the same diet point out the ability of human flora to increase the activity of enteroendocrine cells both in the jejunum and colon. The differences between the effects of the two floras on the distribution of enteroendocrine cells presented in this study therefore emphasise the value of this rat model to determine if different microbial species are involved in the modulation of enteroendocrine cells.

In conclusion, the data presented provide strong evidence that dietary and microbial luminal factors modify the enteroendocrine cells in the intestinal epithelium. However, the molecular mechanisms which control the differentiation of endocrine cells expressing different gastroenteric peptides remain to be resolved.

5 Peyer's Patches of SCID Mice After Syngeneic Normal Bone Marrow Transplantation

The gut-associated lymphoid tissue (GALT), which is present in all vertebrates as ileal Peyer's patches, consists of tightly packed follicles that are separated by small T cell areas and contain 95% surface IgM-positive B cells (Reynaud et al. 1991). Antigens and microorganisms in the intestinal lumen are separated from the lymphoid cells of Peyer's patches by follicle-associated epithelium (FAE) containing M cells (reviewed by Neutra et al. 1996a). M cells are clearly different from neighbouring enterocytes and goblet cells and are specialized for transepithelial transport of antigens and microorganisms (Giannasca and Neutra 1994). The apical membranes of M cells contain glycoconjugates that may mediate the binding of lectins and lectin-like microbial surface proteins (Gebert and Hach 1993). Recent lectin binding studies have identified fucosylated glycoconjugates unique to the apical surfaces and cytoplasmic contents of mouse M cells (Clark et al. 1993; Falk et al. 1994).

The severe combined immunodeficiency (SCID) disease impairs lymphopoiesis and consequently the individuals with this defect are deficient in immune functions mediated by T and B lymphocytes. Mice homozygous for this mutation (SCID mice) are devoid of pre-B, B and T lymphocytes (Reichman et al. 1993) and the lymphoid follicles of the intestinal tract are rarely visible (Bosma et al. 1983).

The experiments described in 3.2.4 have shown that diet and microbial flora can influence the lectin binding pattern of gut epithelial cells. Since M cells in mice are characterised by a particular lectin binding pattern, these cells provide a unique opportunity to study the influence of cell to cell interactions on the glycosylation pattern in the gut. Since SCID mice are devoid of M cells, their reconstitution with normal syngeneic bone marrow should result in the reappearance of Peyer's patches and M cells within these animals. If M cells within the Peyer's patches show the same lectin binding pattern as in the normal BALB/c mice, from which the SCID mouse originates, it would indicate that lymphoid cell to epithelial cell interaction has an influence on the glycosylation of epithelial cells lining the gut lumen. This would add another facet to the regulatory mechanisms of glycosylation in the gut epithelium.

5.1 Results

In all the SCID mice reconstituted with bone marrow progenitor cells from normal syngeneic Balb/c mice, gut-associated lymphoid tissue was identified as single primary follicles or as aggregated Peyer's patches in the terminal ileum. The number of single follicles and Peyer's patches in Balb/c control mice and in SCID mice killed 2, 4,

	Balb/c mice	Bone marrow reconstituted SCID mice			
		2 weeks	4 weeks	6 weeks	10 weeks
Peyers patches	1.58 ± 0.51 (76%)	1.6 ± 0.54 (26%)	1.0 ± 1.2 (12%)	1.2 ± 0.83 (29%)	5.33 ± 0.86[a] (72%)
Single primary follicles	0.5 ± 0.52 (24%)	4.6 ± 0.89[a] (74%)	7.8 ± 2.16[a] (88%)	3.0 ± 1.0[a] (71%)	2.0 ± 0.45[b] (27%)

* % distribution is the percentage of aggregated Peyers patches or single follicles of all follicles present.

[a] Significantly different at $P < 0.001$, ANOVA from controls.

[b] Significantly different at $P < 0.05$ ANOVA from controls.

Fig. 17. Mean number (±SD) and percentage distribution of Peyer's patches and primary follicles in SCID mice and in controls. Percentage distribution is the percentage of aggregated Peyer's patches or single follicles of all follicles present.
[a] Significantly different at $P<0.001$, ANOVA from controls.
[b] Significantly different at $P<0.05$, ANOVA from controls

6 and 10 weeks after bone marrow cell transfer are shown in Fig. 17. Although no significant differences in the number of Peyer's patches were found between the Balb/c and reconstituted SCID mice at 2, 4 and 6 weeks after the bone marrow reconstitution, the number of Peyer's patches at 10 weeks after the bone marrow transfer was significantly higher in reconstituted SCID mice compared with the Balb/c control mice. In reconstituted SCID mice, the number of single follicles was significantly higher at 2, 4, 6 and 10 weeks when compared with the Balb/c control mice.

The FAE covering the domes of the Peyer's patches of SCID mice was composed of enterocytes, goblet cells and M cells, while the FAE overlying the single follicles lacked goblet cells. In the present study, AAA (Fig. 18a–c) and UEA-I (Fig. 18d,e) selectively stained the cell membranes and cytoplasmic contents of M cells in the FAE of the single primary follicles and Peyer's patches of bone marrow reconstituted SCID mice and Balb/c controls. The adjacent enterocytes and goblet cells on FAE were not stained with these lectins. Other lectins, including those with specificities for mannose (Con A) or *N*-acetylgalactosamine (DBA) or sialic acid (MAA), did not bind to M cells. In all sections examined, Con A (Fig. 18f) and MAA (Fig. 18g) were bound to the subepithelial connective tissue while DBA binding was seen in the goblet cells (Fig. 18h).

Preincubations of UEA-I and AAA with fucose resulted in a complete inhibition of their binding to M cells. In control sections incubated with biotinylated Con A, DBA and MAA in the presence of specific monosaccharides, a complete loss of their binding to connective tissue and goblet cells was found.

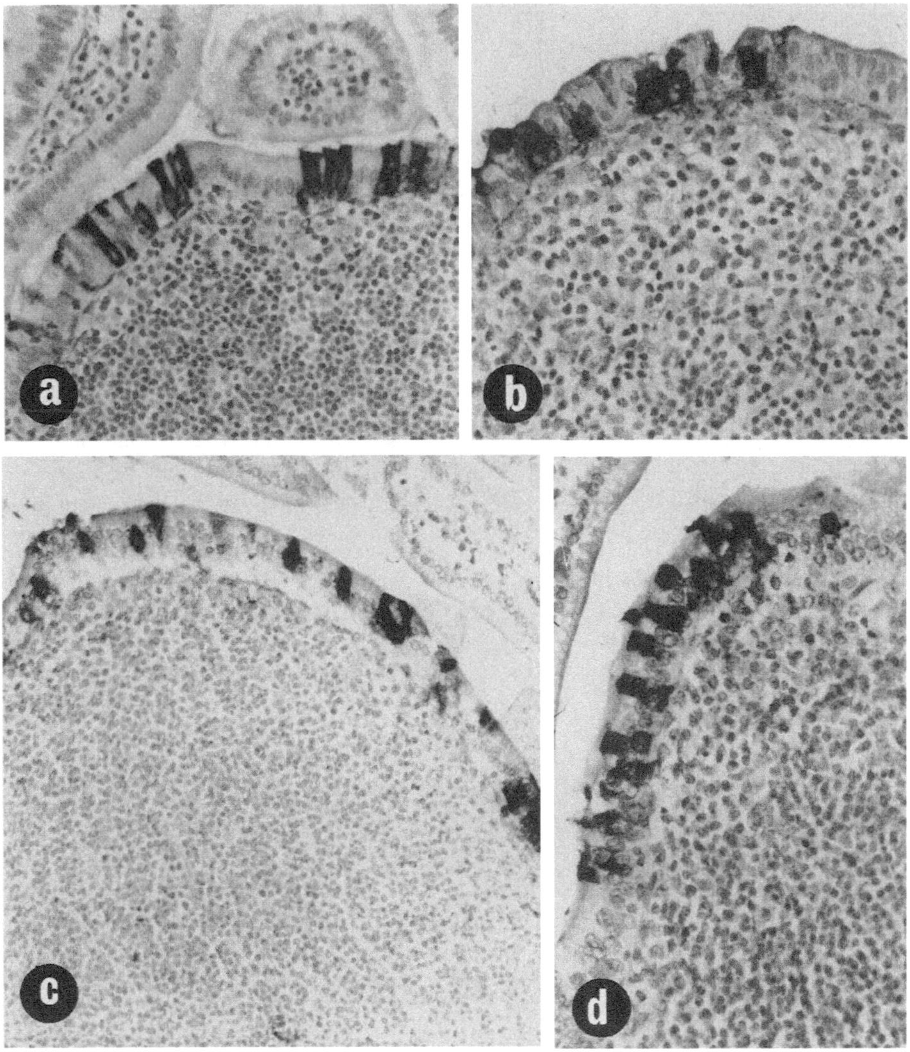

Fig. 18a–d. Glycoconjugate expression of the gut-associated lymphoid tissue in mice. **a** The FAE overlying the single primary follicle of a control Balb/c mouse. The cytoplasm of M cells shows binding sites for *Anguilla* (AAA). ×330. **b** Peyer's patch of a SCID mouse killed 10 weeks after bone marrow reconstitution showing the distribution of *Anguilla* (AAA) binding sites. Population of *Anguilla* stained FAE cells are M cells. ×330. **c** *Anguilla* (AAA) labelling of a single primary follicle in a 10-week bone marrow reconstituted SCID mouse. M cells are selectively labelled. ×250. **d** Single primary follicle of a control Balb/c mouse stained with UEA-I. Cells identified as M cells are strongly labelled. ×250.

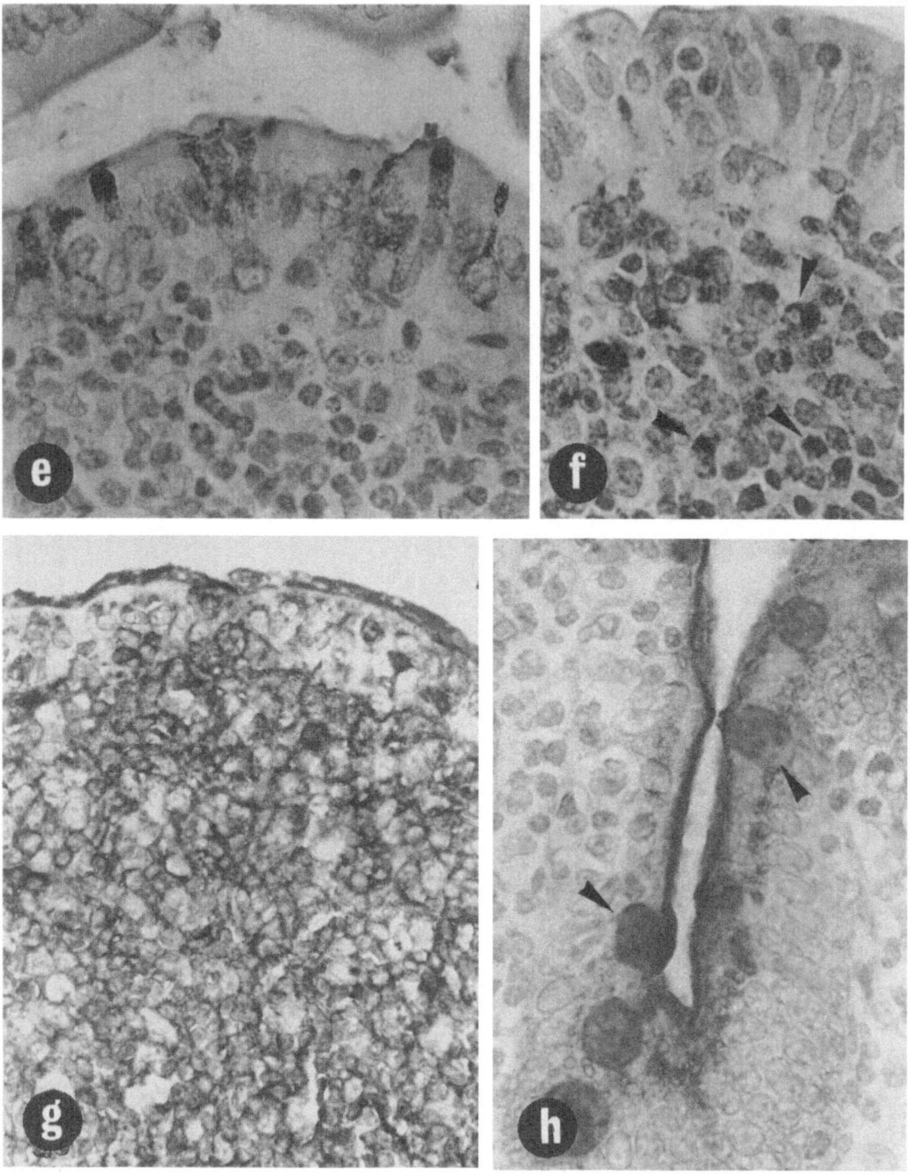

Fig. 18e–h. e UEA-I binding pattern of a Peyer's patch in a SCID mouse killed 10 weeks after bone marrow transfer. Binding is limited to M cells in the FAE. ×500. **f** Con A labelling of a Peyer's patch in a SCID mouse killed 4 weeks after bone marrow transfer from a syngeneic mouse. M cells are not labelled while the subepithelial connective tissue cells (*arrows*) show lectin binding. ×500. **g** MAA binding in a single primary follicle from a SCID mouse killed 6 weeks after bone marrow reconstitution. The connective tissue in the follicle is strongly labelled. ×500. **h** DBA binding sites in a Peyer's patch and adjacent intestinal mucosa from a SCID mouse killed 10 weeks after the bone marrow cell transfer. The goblet cell mucin in crypts is labelled (*arrows*) whereas the follicle-associated epithelium (FAE) is negative. ×500

5.2
Discussion

The development of gut-associated lymphoid tissue depends on the presence of T and B lymphocytes and therefore SCID mice are deficient in Peyer's patches (Bosma et al. 1983; Dorshkind et al. 1984). SCID mice harbour a defective recombinase activity which results in a marked depletion of functional T and B cells during lymphocyte development (Kirchgessner et al. 1995). In contrast, T-cell-deficient nude mice have immature T lymphocytes in the follicle dome beneath the epithelium (Ermak and Owen 1987) and pathogen-free mice have a reduced number of Peyer's patches even though they are fully immunocompetent (Neutra et al. 1996b). Adoptive transfer of syngeneic Peyer's patch and lymph node cells to SCID animals results in extensive infiltration of the lamina propria and the intraepithelial compartments by T cells (Hilbert et al. 1994). However, selective immunoreconstitution of SCID mice with either purified T- or B-cell populations has shown that the B cells are the principal cell lineage responsible for the differentiation of M cells (Savidge 1996).

In the present study, when the percentage distribution and the relative proportion of primary follicles and Peyer's patches in reconstituted SCID mice were compared with the Balb/c controls, there was an increase in the percentage of primary follicles and a decrease in the percentage of Peyer's patches between 2 and 6 weeks after the bone marrow transfer. However, after 10 weeks there was no difference between the reconstituted SCID mice and controls. These data suggest that following transfer of syngeneic bone marrow into SCID mice, lymphocytes initially infiltrate the gut lamina propria to home the single primary follicles. The reconstitution of mucosal tissue by the donor population is well established in Peyer's patches by 10 weeks of the cell transfer.

The M cells characteristically display short, irregular microvilli and basal intraepithelial pockets containing lymphocytes and macrophages (Bye et al. 1984). The apical membranes of M cells contain glycoconjugates that can provide binding sites for microorganisms, and lectin-like surface adhesins of enteric bacteria (Giannasca and Neutra 1994). Previous lectin binding studies have revealed that in the mouse Peyer's patch FAE, α-l-fucose-specific lectins bind selectively to cell membranes and cytoplasmic contents of M cells (Giannasca et al. 1994). The presence of the fucose-specific AAA and UEA-I binding sites to M cells in the present study would imply that the de novo assembly of Peyer's patches by bone marrow reconstitution in SCID mice is accompanied by the development of a dome epithelium with M cells. Hence the contact of the immunocompetent cells with the gut epithelium induces changes in the glycosylation of FAE overlying the lymphoid follicles. This results in the presence of fucose residues on the apical surface and in the cytoplasm of M cells detected by the two fucose-specific lectins used in this study. At present, however, it is not clear whether these fucose residues are derived from an induction of the fucosyltransferases in the epithelium or a downregulation of glycosyltransferases which add terminal carbohydrate moieties masking the binding sites for the two fucose-specific lectins. Irrespective of the mechanisms involved in the expression of fucose residues in M cells, this study proves that lymphocyte–epithelial interactions result in a change in glycosylation of FAE cells. This finding adds a new facet to the role of carbohydrate residues in cell to cell interactions involved in the differentiation of M cells and warrants further investigation.

6 Carbohydrate Expression in Human and Mouse Peyer's Patches

Subepithelial lymphoid follicles are the major site of antigen sampling in the gut and provide an effector mechanism for secretory immune responses in gut-associated lymphoid tissue (GALT) in the intestine (Roy et al. 1987). The generation of an immune response is important for the protection of host tissues against invasion of microorganisms and for limiting infection (Kraehenbuhl and Neutra 1992). Antigens and macromolecules are transported across the follicle-associated epithelium (FAE) overlying the lymphoid follicles which is composed of enterocytes, goblet cells, membranous (M) cells, pre-M-cells and lymphocytes (Bye et al. 1984; Gebert and Hach 1993). Sampling, endocytosis and transport of luminal antigens is accomplished by M cells which are structurally and biochemically specialised to carry out these functions. M cells can be distinguished from enterocytes by the presence of shorter, fewer, widely spaced microvilli and an intraepithelial pocket containing lymphocytes and macrophages (reviewed by Owen and Ermak 1990; Kraehenbuhl and Neutra 1992).

Since some microorganisms and some of the ingested antigens display a selectivity to the FAE and because M cells are initial sites of antigen entry, several studies in recent years have focused on the apical membrane glycoconjugates of M cells that may serve as binding sites for lectins and lectin-like microbial surface proteins (Kraehenbuhl and Neutra 1992; Clark et al. 1993; Gebert and Hach 1993). The identification of glycoconjugate expression and molecules unique to FAE and M-cell surfaces is now considered an attractive experimental model system to develop mucosal vaccines that can be targeted to M cells (Kraehenbuhl and Neutra 1992; Giannasca et al. 1994; Owen 1994).

The GALT and FAE have been the subject of numerous immunocytochemical and lectin histochemical studies in mice (Bye et al. 1984; Clark et al. 1993; Giannasca et al. 1994), rats (Owen and Bhalla 1983; Allan et al. 1993) and rabbits (Neutra et al. 1987; Roy and Varvayanis 1987; Gebert and Schumacher 1991; Gebert and Hach 1993). However, no studies on the GALT and FAE associated glycoconjugates of human Peyer's patches have been reported. The aim of the present study was to investigate the binding of lectins, with emphasis on recently isolated lectins, to the human GALT and the overlying FAE in order to identify their cell-associated carbohydrates. To determine whether the glycoconjugate expression of human GALT and FAE differs between animal species, the lectin binding pattern on corresponding tissues taken from the BALB/c mouse with those of the human Peyer's patches was compared. This comparison will answer the question whether the expression of carbohydrate residues in mice and human M cells are the same. This result has bearings on future experimental design in which the glycosylation of M cells plays a crucial role such as vaccination studies using lectins for the targeting of M cells.

6.1
Results

The cellular pattern of lectin binding in the FAE, villus epithelium and GALT of mouse and human Peyer's patches are summarised in Tables 9 and 10. In control sections incubated in the presence of specific inhibitory monosaccharides, the lectin binding was inhibited or greatly diminished after this absorption. The major findings with the lectin panel and the staining differences between the corresponding cells in the mouse and human intestinal tissues are highlighted below.

6.1.1
α-Fucose-Specific Lectins

In the FAE of the mouse, UEA-I and AAA (Fig. 19a,b) selectively stained the cell membranes and cytoplasmic contents of M cells. The enterocytes and goblet cells were not stained with these lectins. In contrast, UEA-I and AAA binding was seen in the epithelial cells of human FAE with the exception of goblet cells; however, no qualitative difference between M cells and non-M cells could be seen. LTA recognised both M cells and other enterocytes of mouse FAE, while no LTA binding was seen in M cells and enterocytes of human FAE. LTA reactivity was observed in the mucin granules of villus epithelium in both the mouse and human. No binding sites for any fucose-specific lectins were observed in cells of GALT.

6.1.2
Mannose-Specific Lectins

Apart from GNA, no staining was detected in the mouse M cells using mannose-specific lectins. Binding of NPA was observed only in the enterocytes of the mouse FAE. The macrophages in the germinal centre of mouse GALT showed an intense binding to AMA and in paracortical areas to NPA, GNA and HHA. In the human GALT, although no NPA reactivity was seen in the macrophages, there were numerous binding sites for AMA, AUA, LOA, GNA and HHA in tingible body macrophages of the germinal centre. Goblet cells both in mouse and human gut epithelium did not react with any of the mannose-specific lectins. A few of the mannose-specific lectins (AMA, LOA, GNA, HHA) labelled M cells and enterocytes with equal intensity in the human FAE.

Table 9. Binding of lectins to mouse Peyer's patches

Lectin specificity	Lectin	FAE			Villus epithelium		GALT	
		M cells	Entero cytes	Goblet cells	Enter ocytes	Goblet cells	Lympho cytes	Macro phages
α-l-Fucose	UEA-I	4	0	0	0	0	0	0
	LTA	2	1	0	0	3	0	0
	AAA	4	0	0	0	0	0	0
Mannose	AMA	0	1	0	2	0	1	3
	ASA	0	0	0	0	0	0	0
	AUA	0	0	0	0	0	0	0
	LOA	1	1	0	1	0	1	0
	NPA	0	3	0	2	0	0	4
	GNA	4	1	0	2	0	2	4
	HHA	0	2	0	1	0	2	3
d-GalNAc/l-fucose	PTA	3	2	3	0	1	0	0
d-GalNAc	SBA	0	1	0	0	2	0	0
	APA	2	0	0	1	2	0	0
	BDA	0	0	0	0	0	4	4
	BPA	0	0	1	0	1	0	0
	DBA	2	2	0	2	0	0	2
	EHA	0	0	0	0	0	0	0
	IRA	1	0	0	0	0	1	2
	ML-II	0	0	0	0	0	0	0
	SNA-II	2	0	2	0	2	0	0
	VVA	0	1	0	2	3	0	0
	HPA	0	0	0	0	3	0	0
d-Gal/GalNAc	ACA	0	0	0	0	0	0	0
	GSA-I	0	1	0	1	0	3	4
d-GlcNAc	CMA	2	2	2	1	2	1	4
	EEA	2	0	0	1	0	3	2
Complex	CAA	0	1	0	2	0	3	4

See Table 4 for lectin abbreviations. *Numbers* indicate staining frequency on a semiquantitative scale ranging from 0 to 4 where 0 corresponds to no reactive cells, 1 to occasional, 2 to a few, 3 to a moderate number and 4 to numerous reactive cells.

Table 10. Binding of lectins to human Peyer's patches

Lectin specificity	Lectin	FAE			Villus epithelium		GALT	
		M cells	Entero cytes	Goblet cells	Enter ocytes	Goblet cells	Lympho cytes	Macro phages
α-l-Fucose	UEA-I	2	1	0	1	0	0	0
	LTA	0	0	1	0	1	0	0
	AAA	2	1	0	1	0	0	0
Mannose	AMA	2	2	0	2	0	1	3
	ASA	0	0	0	0	0	0	0
	AUA	0	0	0	0	0	0	4
	LOA	2	2	0	1	0	0	3
	NPA	0	0	0	0	0	0	0
	GNA	2	2	0	1	0	0	4
	HHA	2	2	0	2	0	0	4
d-GalNAc/l-fucose	PTA	2	2	3	2	3	0	0
d-GalNAc	SBA	0	0	0	0	0	0	0
	APA	0	0	0	2	0	0	0
	BDA	0	0	1	0	1	4	4
	BPA	2	2	0	2	0	0	4
	DBA	2	2	2	1	3	0	2
	EHA	0	0	0	0	0	0	0
	IRA	0	0	0	1	0	0	4
	ML-II	3	3	1	0	2	0	0
	SNA-II	1	1	2	0	3	0	3
	VVA	0	0	2	1	4	0	0
	HPA	3	3	2	1	4	0	0
d-Gal/GalNAc	GSA-I	0	0	0	0	1	0	0
	ACA	0	0	0	0	0	0	0
d-GlcNAc	CMA	0	0	0	0	0	4	0
	EEA	2	2	0	1	0	0	0
Complex	CAA	0	0	0	0	0	0	4

See Table 4 for lectin abbreviations. Numbers indicate staining frequency on a semiquantitative scale ranging from 0 to 4 where 0 corresponds to no reactive cells, 1 to occasional, 2 to a few, 3 to a moderate number and 4 to numerous reactive cells.

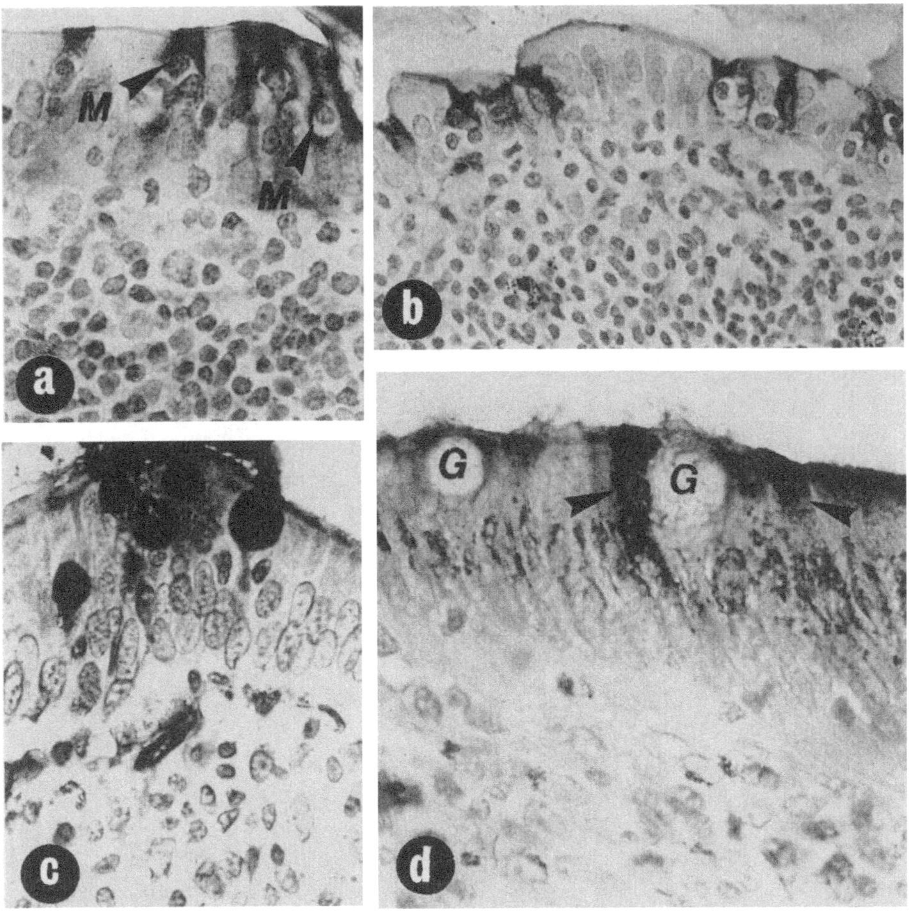

Fig. 19a–d. Lectin binding sites in the mouse and human intestinal mucosa. **a** UEA binding sites in mouse FAE. M cells (*M*) containing intraepithelial lymphocytes are selectively labelled. ×660. **b** The FAE overlying a dome of the mouse lymphoid follicle showing the distribution of AAA sites. The cytoplasm of M cells is selectively labelled. Non-FAE cells are devoid of labelled cells. ×330. **c** Human Peyer's patch stained with PTA. The goblet cells of FAE and mucus are heavily labelled. ×660. **d** Labelling of human FAE with HPA. HPA binds to the FAE to a variable extent; goblet cells (*G*) are not labelled while enterocytes are predominantly apically labelled; cells presumably representing M cells (*arrows*) are labelled by HPA throughout the entire cytoplasm. However, no distinct lectin binding pattern of M cells can be established (compare with **a** and **b** in mice). ×660.

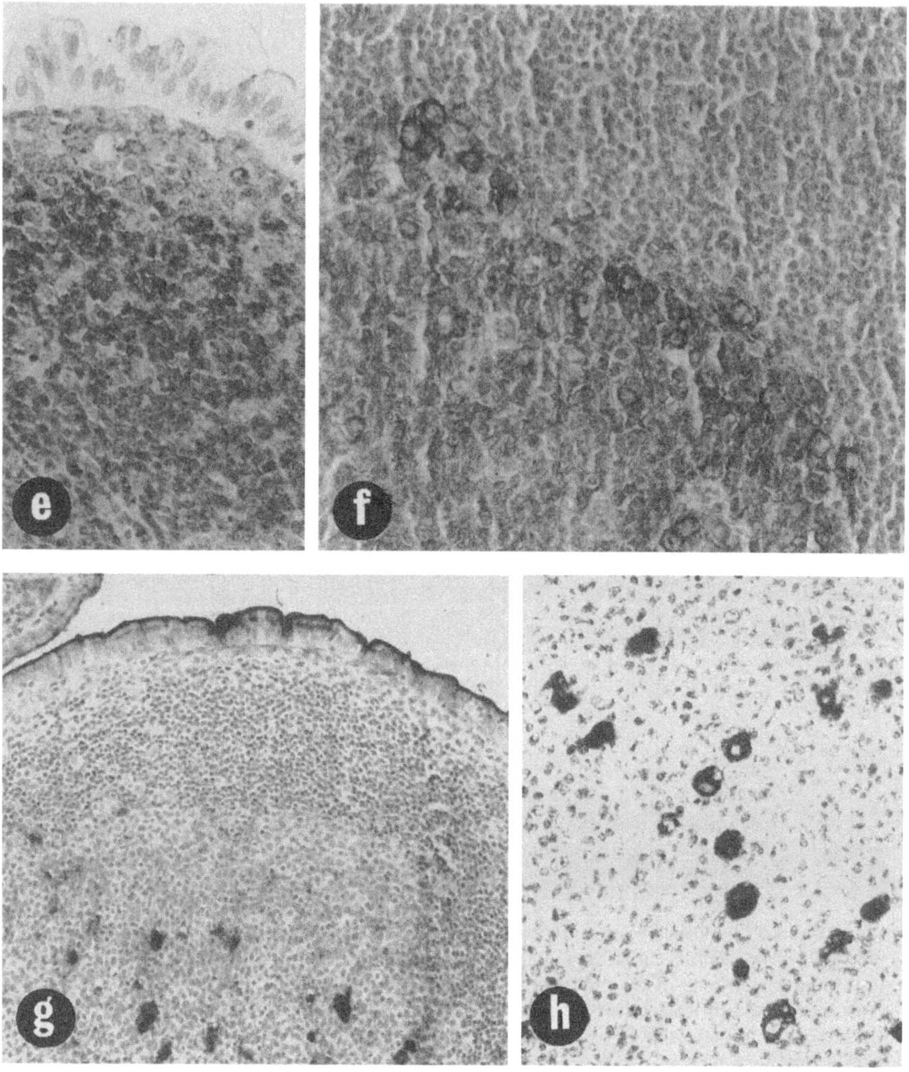

Fig. 19e–h. e Mouse intestinal lymphoid follicles labelled with BDA. The lymphocytes in the paracortical region are strongly labelled. ×165. **f** Labelling of human Peyer's patch with CMA. The germinal centre lymphocytes and macrophages are strongly labelled. ×330. **g** In mouse lymphoid follicle, CMA produces prominent labelling of germinal centre macrophages. ×165. **h** Human Peyer's patch stained with CAA. Cells identified as germinal centre tingible body macrophages are strongly labelled. ×330

6.1.3
GalNAc-Fucose-Specific Lectin-PTA

PTA stained M cells and goblet cells more strongly than other epithelial cells in the mouse. PTA reactivity was strongest in the M cells in the mouse, while in the human intestine the staining was most intense in the goblet cells (Fig. 19c).

6.1.4
GalNAc-Specific Lectins

ML-II and HPA (Fig. 19d) labelled both enterocytes and presumed M cells of human FAE. In the human intestine, the GalNAc-specific lectins (DBA, SNA-II, VVA and HPA) stained goblet cells more strongly in the villus crypts than the goblet cells in the FAE. BDA reacted strongly with both lymphocytes and macrophages of paracortical area of both human and mouse (Fig. 19e) intestine. BPA and IRA bound intensely to the tingible body macrophages of human GALT but there was no staining of lymphocytes. There were no binding sites for SBA and EHA in the human ileum and for MLII in the mouse intestine.

6.1.5
Gal/GalNAc-Specific Lectins

There was no staining using ACA. No staining was detected using GSA-I in human ileum apart from occasional positivity in goblet cells in the villi. However, binding sites for GSA-I were observed in all of mouse GALT lymphocytes and germinal centre macrophages.

6.1.6
GlcNAc-Specific Lectins

CMA and EEA stained M cells in the mouse and there was EEA binding in non-goblet cells of human FAE. Significant CMA binding was observed only in germinal centre B lymphocytes of human GALT (Fig. 19f) and in tingible body macrophages of mouse GALT (Fig. 19g). EEA binding sites were not seen in human GALT, but were strongly expressed in all the lymphocytes of the mouse GALT.

6.1.7
Complex Lectin-CAA

CAA reactivity was detected in enterocytes of mouse FAE and villus epithelium. CAA did not react with the human FAE and villus epithelium, although it did stain the GALT tingible body macrophages (Fig. 19h). CAA binding sites were observed in the paracortical lymphocytes and tingible body macrophages in the mouse.

6.2 Discussion

Although lectin histochemical studies have been used extensively to investigate the distribution of glycoconjugates in the FAE and GALT of intestinal mucosa in various animal species, including mouse (Clark et al. 1993; Falk et al. 1994; Giannasca et al. 1994), rat (Owen and Bhalla 1983) and rabbit (Gebert and Hach 1993; Jepson et al. 1995), no data about the lectin binding pattern of human FAE and GALT have been reported. The present study was undertaken to define accurately the distribution of binding sites for a range of lectins directed against various different sugar residues in human and mouse intestine. In addition to the lectins used by previous authors, this investigation employed new lectins isolated and characterised by Peumans and Van Damme (1994), who kindly supplied the lectins for this study (see Table 5). These lectins have not previously been used in the study of glycoconjugate distribution of intestinal tissues. The results indicate that marked inter-species differences in the glycoconjugate expression exist between the constituent cells within the FAE and GALT of mouse and human intestinal mucosa.

The distinctive profile of UEA-I and AAA-staining in mouse M cells in the present study confirmed the data of Giannasca et al. (1994) who described the presence of fucose residues in the M cells of mouse Peyer's patches. The immunocytochemical and ultrastructural characteristics of human M cells have previously been reported (Owen and Jones 1974; Nagura et al. 1991; Cuvelier et al. 1994), but the lectin binding pattern of human FAE has not yet been described. While four of the lectins used in this study were able to distinguish the M cells from the enterocytes within the mouse FAE, no such differential expression of lectin binding sites was observed in human FAE. M cells are known to be scarce in normal human ileal mucosa and constitute 0%–6% of the cells in FAE whereas in inflamed mucosa their proportion increases up to 24% (Cuvelier et al. 1994). The scarcity of M cells in human FAE taken together with the fact that they do not have a distinct lectin binding pattern would suggest that lectins may not serve as useful markers for normal human M cells or as a vehicle for vaccination studies.

The present study also demonstrated that the FAE enterocytes and goblet cells possess lectin-binding characteristics distinct from the villus enterocytes and goblet cells. The differences were more prominent in the human FAE where two of the mannose-specific lectins (LOA and GNA) and four of the GalNAc-specific lectins (DBA, ML-II, SNA and HPA) recognised a larger population of FAE enterocytes. In contrast, DBA, ML-II, SNA-II, VVA and HPA that recognise GalNAc residues showed restricted labelling of FAE goblet cells. These findings support the results of Bhalla and Owen (1982) who showed that the goblet cell activity is suppressed in mouse FAE and the recent observations of Jepson et al. (1995) who showed that FAE enterocytes display a higher affinity for PNA in rabbit small intestine.

Several of the lectins displayed distinctive patterns of binding to B lymphocytes in the nodules, internodular T-cell areas and macrophages of the GALT. For example, in the mouse lymphoid follicles BDA, GSA-I, EEA and CAA labelled both B and T lymphocytes, but in human Peyer's patches BDA selectively labelled paracortical T lymphocytes and CMA labelled germinal centre B cells. A heterogeneous population of macrophages was identified in this study on the basis of their glycoconjugate expression. The tingible body macrophages (Kroese et al. 1990) that reacted with mannose and GalNac-specific lectins in human intestine were restricted to the germinal centres, whereas in the mouse these lectins also labelled the paracortical macrophages in the T-cell area. The germinal centre tingible body macrophages in the mouse showed a unique lectin profile and were labelled by GSA-I that recognises Gal/GalNAc, by CMA that recognises GlcNAc and by CAA that recognises complex type carbohydrate residues.

In conclusion, based on the reactivity of a comprehensive panel of lectins which were specific to the anomeric linkage of the glycoconjugate side chains, this study suggests that terminal fucose residues that bound via α (1,2) linkages to galactose occur exclusively in M cells of the mouse FAE, whereas in human M cells, only α- or α-linked GalNAc residues are present. Although the precise mechanisms involved in the adherence of microorganisms to gut luminal surface are yet unknown, interactions between the lectin-like substances on the bacterial surface and M-cell membrane glycans have been shown to facilitate the attachment of pathogenic bacteria to the intestinal epithelium (Sharon 1987). The demonstration of differences in the glycoconjugate residues between M cells, enterocytes and GALT of the mouse and human intestinal mucosa in this study reflect species-specific glycoconjugate expression which might partly represent adaptation of different microbial populations. These differences have to be born in mind when targeting these cell populations.

7 General Discussion and Conclusions

The vast mucosal surface of the digestive tract is lined by a single-layered epithelium. The epithelial cells not only allow passage of nutrients but are also uniquely equipped to face a foreign environment rich in potential pathogens. Many antigens and microorganisms are prevented from contact with epithelial surface by local secretions, especially mucins and secretory IgA antibodies. The major experiments conducted were specifically aimed to investigate how the luminal interactions of the diet and microflora in a rat model modify the morphology of villi and crypts, alter the carbohydrate expression of mucins and influence the activity of enteroendocrine cells of the intestinal mucosa. The advantage of using an animal model is that it allows the regulation of diet and experimental manipulation of bacterial flora that would be unacceptable or impossible in a human subject. Two diets differing widely in characteristics such as texture, composition, digestibility etc. were chosen to enhance the possibility of detecting alterations in gut microstructure and epithelial mucins. The knowledge of the dietary mechanisms which alter the biosynthesis, turnover and degradation of intestinal mucins from this study are of value in understanding both the normal gut ecosystem and various intestinal disorders. Any information derived from this study may therefore also be relevant even in management of such diseases. This study also focused on the intestinal epithelium, FAE, overlying the gut-associated lymphoid tissue (GALT). In line with the investigations on mucins, the emphasis was placed on glycoconjugates of M cells which are relevant to targeting of vaccines to the mucosal immune system.

The main and most important finding of this work is that the dietary constituents, microbial flora and their interactions in the gut lumen modify the mucus composition of intestinal mucosa as a result of alteration in synthetic/secretory activity of its goblet cells. A number of experimental techniques were employed to obtain the evidence for these changes. The use of quantitative morphological methods reveals that feeding of a purified diet results in shallower crypts in the small intestine but deeper crypts in the large intestine when compared with a fibre-rich commercial diet. A more specific alteration in distribution of neutral mucins occurs in animals fed a commercial diet which are more intense in the upper crypts of small intestine than in the lower crypts, and to a lower extent in the upper crypts of the large intestine. The experimental data derived from morphometric studies provide further evidence of quantitative changes in the number, volume density and staining density of stored mucins in the goblet and deep crypt cells of animals fed different diets and exposed to different microflora. Feeding a commercial diet reduced the volume density of cells containing neutral and sulphomucins in the jejunum of conventional rats and the staining density of neutral and acidic mucins in the germ-free rats. The association between diet and mucin composition is further supported by local and regional changes in goblet cell glycoconjugates revealed by lectin histochemical studies. The studies clearly demonstrate the enhancing effect of nondigestible dietary fibre in the commercial diet on the luminal aspect of the small intestine where the surface goblet cells were found intensely labelled with *N*-acetyl-glucosamine and sialic acid-linked d-galactose-specific lectins. A similar increased staining with an *N*-acetylgalactosamine-specific lectin induced by a commercial diet in the colon suggests that the fibrous component in the diet enhances intestine protection. On the basis of these results, it is proposed that dietary manipulation modifies the crypt-villus architecture, mucin composition and carbohydrate residues of goblet cell glycoconjugates, leading to alterations in the functional characteristics of the intestinal tract.

The data presented in this study show the ability of enteric flora to modulate the mucin secretion and reveal how the microbial interactions with different diets modify the histochemical composition of goblet cell mucins. As shown by quantitative data, the human flora specifically reduces the villus and crypt dimensions in both the small and large intestine. Human flora also has a significant effect on the mucin secretory activity of goblet cells; it reduces the amount of neutral mucins in the goblet cells of both small and large intestine and increases the amount of sulphated mucins in the large intestine. When the quantitative data on stored mucin was analysed, both rat and human floras were found to reduce the number of cells containing acidic and sulphomucins and the staining density of neutral mucins in the small intestine of animals fed on a purified diet. However, inoculation of human flora increased the staining density of stored neutral and sulphated mucins in the cells of the large intestine. More specific changes in terminal carbohydrate residues of goblet cell mucin are revealed by lectin histochemical analysis which clearly show how intestinal mucosa adapts to different microbial populations. Two species of microflora tested in this study were capable of modifying selected carbohydrate terminal structures associated with mucin glycosylation. An overall reduction in sialic acid-linked d-galactose residues in conventional rats and a loss of crypt-to-surface gradient of fucosyl expression in the large intestine of human flora rats are likely to be due to differing strains of glycosidases in the two microflora.

The results of the experiments in this study provide the first evidence for an influence of diet and bacteria on carbohydrate composition of mucins. The changes in the amount and proportion of mucins after microbial challenge are more dramatic in the colon whereas the dietary manipulation is more influential in modifying the mucin secretion in the small intestine. This reflects the microbiological finding that bacterial colonisation is much more prominent in the colon than in the jejunum. To test the hypothesis whether alterations in carbohydrate composition of mucins are due to changes in mucin oligosaccharide biosynthesis and degradation, one has to analyse the glycosyltransferase and glycosidase enzyme activities in the animals. Although there are no histological methods available to demonstrate the glycolytic activity of bacteria, the activity of various glycosidases can be measured by the rate of release of *p*-nitrophenol from the substrate solution of *p*-nitrophenolglycoside and mucosal contents as described by Meslin et al. (1993) and Corfield et al. (1993). Clearly, these approaches will be useful in future experiments to determine the changes in the quantity and nature of these enzymes.

Several functions of the gut are locally influenced by peptides and biogenic amines released from enteroendocrine cells. This study evaluates the hypothesis whether the feeding of different diets and an association of different microflora modifies the secretion of peptide hormones. The results of quantitation of argyrophil and serotonin immunoreactive cells along the crypt-surface and cephalo-caudal axes of the intestine clearly indicate that there are differences in the number of enteroendocrine cells which are due to the two different diets and floras. Three main findings emerge from these investigations: (1) feeding the commercial diet reduces the number of argyrophil endocrine cells in the jejunum and serotonin immunoreactive cells in the colon of germ-free animals but increases the serotonin immunoreactive cells in the colon of conventional animals; (2) the rat flora increases the serotonin immunoreactive cells in the colon of animals fed a commercial diet and decreases in those fed a purified diet; and (3) inoculation of human flora increases the numbers of serotonin immunoreactive cells both in jejunum and colon. The results provide evidence how the diet-microbial interactions can alter the regional number of enteroendocrine cells and therefore, alter the endocrine status of the intestinal tract.

An experimental SCID mouse model was designed to investigate the histogenesis of Peyer's patches and dynamics of the M-cell development. The quantitative developmental study described in Sect. 5 demonstrates that the single (primary) follicles and aggregates of lymphoid nodules, the Peyer's patches, are generated in SCID mice by reconstitution of these mice with bone marrow cells from syngeneic mice. To test the hypothesis that repopulated T and B cells in Peyer's patches can induce generation of M cells, the study employed two fucose-specific lectins, *Anguilla anguilla* (AAA) and *Ulex europaeus* I (UEA-I), previously shown as positive markers of mouse Peyer's patch M cells. The results of these experiments reveal that M cells develop in the dome epithelium overlying the single primary follicles and Peyer's patches of reconstituted SCID mice.

These data indicate that not only diet and microbial flora can alter glycoconjugate expression in the gut, but that cells immigrating into the gut from the abluminal site have an influence on the glycosylation of the cells lining the gut lumen as well.

The objectives of the studies described in Sect. 6 of this work were to identify lectins which may discriminate among Peyer's patch populations in mice and humans by exploiting differences in their glycoconjugate expression. Most previous studies have used mice as models for human disease to detect glycoconjugates which could be exploited for antigen delivery and triggering local immune responses. The hypothesis that UEA-I and AAA lectins could be considered as good candidates to design strategies of mucosal vaccination represents an extrapolation but was never tested in human. This is the first study to analyse the glycoconjugate distribution of human Peyer's patches. The results clearly indicate that fucosyl glycoconjugates expressed by M cells in mouse do not bind specifically to human M cells, excluding the use of the lectins UEA-I and AAA as delivery vectors to human Peyer's patches. Furthermore, the reaction patterns obtained by recently characterised lectins reveal distinct differences in glycosylation between mouse and human Peyer's patches and their associated lymphoid cells. For example, in the human Peyer's patches, *Bryonia dioica* lectin bound selectively to paracortical T lymphocytes and *Chelidonium majus* lectin to germinal centre B cells. Certain mannose-specific lectins (*Galanthus nivalis*, *Hippeastrum hybrid*) stained the tingible body macrophages in the germinal centre of human Peyer's patches but labelled the macrophages in the paracortical T-cell region of the mouse. It should be stressed that when considering cell surface glycoconjugates as target molecules for the gut immune system, care has to be taken to choose the appropriate lectin for each species.

7.1 Summary

The understanding of mechanisms responsible for alterations in mucin synthesis and secretions is essential in elucidating the aetiology of intestinal disorders. The existence of distinct mucins and M-cell transport mechanisms, and their beneficial effects, have long been recognised. Since nutritional and bacterial factors alter the mucin characteristics and are relevant to planning preventative strategies for intestinal diseases, the studies described in this monograph were designed to test the hypothesis that mucin composition and endocrine status of the intestinal tract are altered by dietary constituents and microbial flora. A study of gut-associated lymphoid tissue (GALT) was undertaken to test the hypothesis that lymphocyte–epithelial interactions influence the glycosylation of cells overlying Peyer's patches.

The effects of diet and microbial flora were analysed by comparing the data from male Wistar germ-free rats, with conventional or human flora. Such rats were fed either a commercial diet, containing crude fibre, or a purified diet. Bone marrow transfers from syngeneic Balb/c mice to severe combined immunodeficient (SCID) mice were made to induce the formation of GALT. A comparison was made of the glycoconjugate profile of Peyer's patches in the guts of mice and humans. The results of the study showed that feeding a fibre-rich diet to germ-free rats resulted in increased villus-crypt lengths, decreased carboxylated mucin content of goblet cells, increased *N*-acetylglucosamine and sialic acid residues in the surface goblet cells, and a reduction in the number of endocrine cells in the small intestine. The lectin markers revealed M-cell development in the dome epithelium of the reconstituted SCID mice. Mouse M cells were labelled by fucose-specific lectins but in human Peyer's patches no distinct M-cell-staining pattern was observed.

Conclusions: (1) the interactions between diet and flora alter the mucosal architecture and the activity of endocrine cells; (2) the dietary changes are influential in modifying the epithelial mucin predominantly in the small intestine while the microbial flora influences the mucosal architecture predominantly in the large intestine; (3) the use of bone marrow transplantation from syngeneic mice into SCID mice along with lectin markers for M cells can be used to study the histogenesis of Peyer's patches; (4) the distinct differences between mouse and human Peyer's patches suggest that when considering cell surface glycoconjugates as target molecules appropriate lectin should be used for each species.

A future challenge in intestinal epithelial cell biology is identifying the nature and distribution of cell surface receptors for specific dietary components and bacteria.

References

Aiken KD, Yu W, Wright JR Jr, Roth KA (1994) Adaptation of enteroendocrine cells in response to jejunal-ileal transposition in the rat. Gastroenterology 106:1576–1583

Allan CH, Mendrick DL, Trier JS (1993) Rat intestinal M cells contain endosomal-lysosomal compartments and express class II major histocompatibility complex determinants. Gastroenterology 104:698–708

Allen A (1978) Structure of gastrointestinal mucus glycoproteins and the viscous and gel-forming properties of mucus. Br Med Bull 34:28–33

Allen A (1981) Structure and function of gastrointestinal mucus. In: Johnson CR, Christensen JM, Grossman I, Jacobson ED, Schultz SG (eds) Physiology of the gastrointestinal tract, 1st edn. Raven, New York, pp 617–639

Allen A, Hutton DA, Leonard AJ, Pearson JP, Sellars LA (1986) The role of mucus in the protection of the gastrointestinal mucosa. Scand J Gastroenterol 21:71–77

Allen A, Pain R, Snary D (1974) Formation of mucous gel from the high molecular weight mucoprotein of gastric mucus. Discussions Faraday Chem Soc 57:210–220

Allen A, Pearson JP (1993) Mucus glycoproteins of the normal gastrointestinal tract. Eur J Gastroenterol Hepatol 15:193–199

Allen A, Snary D (1972) The structure and function of gastric mucus. Gut 13:666–672

Allen A, Starkey BJ (1974) Neuraminidase in pig gastric mucus. Biochim Biophys Acta 338:364–368

Altmann GG (1971) Influence of bile and pancreatic secretion on the size of the intestinal villi in the rat. Am J Anat 132:167–178

Altmann GG (1983) Morphological observations on mucus-secreting non goblet cells in the deep crypts of ascending colon. Am J Anat 167:95–117

Anderson RGW, Kaplan J (1983) Receptor-mediated endocytosis. In: Satir B (ed) Modern cell biology. Liss, New York, pp 1–52

Audie JP, Janin A, Porchet MC, Gosselin B, Aubert JP (1993) Expression of human mucin genes in respiratory, digestive, and reproductive tracts ascertained by in situ hybridization. J Histochem Cytochem 41:1479–1485

Bara J, Loisillier F, Burtin P (1980) Antigens of gastric and intestinal mucous cells in human colonic tumours. Br J Cancer 41:209–221

Barber DL (1993) Regulation of peptide secretion from gastroenteric endocrine cells. In: Brown DR (ed) Gastrointestinal regulatory peptides. Springer, Berlin Heidelberg New York, pp 105–131

Barber DL, Cacace AM, Raucci DT, Ganz MB (1991) Fatty acids stereospecifically stimulate neurotensin release and increase $[Ca2+]^i$ in enteric endocrine cells. Am J Physiol 261:G497–G503

Bhalla DK, Owen RL (1982) Cell renewal and migration in lymphoid follicles of Peyer's patches and caecum – an autoradiographic study in mice. Gastroenterology 82:232–242

Bird GW (1959) Agar gel studies of blood group specific substances and pecipitins of plant origin. II. The precipitins of *Ricinus communis*. Vox Sanguinis 4:313–317

Boland CR, Ahnen DJ (1985) The binding of lectins to goblet cell mucin in premalignant colonic epithelium of the CF-1 mouse. Gastroenterology 89:127–137

Boland CR, Lance P, Levin B, Riddell RH, Kim YS (1982) Lectin binding indicates an abnormality of the goblet cell glycoconjugates in ulcerative colitis. Gastroenterology 82:1021A

Boland CR, Montgomery CK, Kim YS (1982a) A cancer-associated mucin alteration in benign colonic polyps. Gastroenterology 82:664–672

Boland CR, Montgomery CK, Kim YS (1982b) Alteration in human colonic mucin occurring with cellular differentiation and malignant transformation. Proc Natl Acad Sci USA 79:2051–2055
Bosma GC, Custer RP, Bosma MJ (1983) A severe combined immunodeficiency mutation in the mouse. Nature 301:527–530
Brackett KA, Townsend SF (1980) Organogenesis of the colon in rats. J Morphol 163:191–201
Brasitus TA, Goldfarb JP, Hsu KC, Tannenbaum M (1982) Characterization of rat colonic cell surface glycoconjugates by fluoresceinated lectins. I. Importance of fixation techniques. Histochemistry 76:197–209
Brown D, Cremaschi D, James PS, Rosetti C, Smith MW (1990) Brush-border membrane alkaline phosphatase activity in mouse Peyer's patch follicle-associated enterocytes. J Physiol (Lond) 427:81–88
Brown R, Kelleher J, Losowsky MS (1979) The effect of pectin on the structure and function of rat intestine. Br J Nutr 42:357–365
Bye WA, Allan CH, Trier JS (1984) Structure, distribution and origin of M cells in Peyer's patches of mouse ileum. Gastroenterology 86:789–801
Carlstedt-Duke B (1989) The normal microflora and mucin. In: Grubb R, Midtvedt T, Norin E (eds) The regulatory and protective role of the normal microflora. Wenner-Gren International Symposium Series, vol 52. Macmillan, London, pp 109–127
Carlstedt I, Sheehan JK (1989) Structure and macromolecular properties of cervical mucus glycoproteins. In: Chantler E, Ratcliffe NA (eds) Mucus and related topics. Symposia of Society of Experimental Biology 43:289. University of Cambridge
Cassidy MM, Lightfoot FG, Grau LE, Story JA, Kritchevsky D, Vahouny GV (1981) Effect of chronic intake of dietary fibres on the ultrastructural topography of rat jejunum and colon: a scanning electron microscopy study. Am J Clin Nutr 34:218–228
Cassidy MM, Lightfoot FG, Vahouny GV (1982) Morphological aspects of dietary fibre in the intestine. Adv Lipid Res 19:203–228
Chadurek VS, Gaginella TS, Carlson GL, Debongnie JC (1979) Effects on molecular structure of bile-acid induced alterations in absorptive function, permeability and morphology. J Lab Clin Med 94:661–664
Chang SK, Dohrman AF, Basbaum CB, Ho SB, Tsuda T, Toribara NW, Gum JR, Kim YS (1994) Localization of mucin (MUC2 and MUC3) messenger RNA and peptide expression in normal human intestine and colon cancer. Gastroenterology 107:28–36
Cheng H, Leblond CP (1974) Origin, differentiation and renewal of the four main epithelial cell types in the mouse small intestine. V. Unitarian theory of the origin of the four epithelial cell types. Am J Anat 141:537–561
Clamp JR, Gough M (1991) The glycocojugate content of human faeces and ileostomy effluents. Clin Chem 203:243–248
Clark MA, Jepson MA, Simmons NL, Booth TA, Hirst BH (1993) Differential expression of lectin-binding sites defines mouse intestinal M-cells. J Histochem Cytochem 11:1679–1687
Clark MA, Jepson MA, Simmons NL, Hirst BH (1994) Differential surface characteristics of M cells from mouse intestinal Peyer's patches and caecal patches. Histochem J 26:271–280
Clark MA, Jepson MA, Simmons NL, Hirst BH (1995) Selective binding and transcytosis of *Ulex europaeus* 1 lectin by mouse Peyer's patch M-cells in vivo. Cell Tissue Res 282:455–461
Clarke RM (1977) "Luminal nutrition" versus "functional work load" as controllers of mucosal morphology and epithelial replacement in rat small intestine. Digestion 15:411–424
Coates ME, Walker R (1992) Interrelationships between the gastrointestinal microflora and non-nutrient components of the diet. Nutr Res Rev 5:85–96
Colacchio TA, Chabot JA, Zimmerman BW (1984) Differential mucin staining in colorectal neoplasms – potential clinical application. Am J Surg 147:666–669
Conway PL, Welin A, Cohen PS (1990) Presence of K88-specific receptor in porcine ileal mucus is age dependent. Infect Immun 58:3178–3182
Corfield AP, Paraskeva C (1993) Secreted mucus glycoproteins in cell and organ culture. In: Hounsell EF (ed) Glycoprotein analysis in biomedicine 14. Humana, Totawa, pp 211–232
Corfield AP, Wagner SA, O'Donnell LJD, Durdey P, Mountford RA, Clamp JR (1993) The roles of enteric bacterial sialidase, sialate *O*-acetyl esterase and glycosulfatase in the degradation of human colonic mucin. Glycoconjugate J 10:72–81

Craig SW, Cebra JJ (1971) Peyer's patches: an enriched source of precursors for IgA producing immunocytes in the rabbit. J Exp Med 134:188–200
Creamer B (1974) Intestinal structure in relation to absorption. Biomembranes 4A:1–42
Cromwell CL, Hoskins LC (1977) Antigen degradation in human colon ecosystems. Host's ABO blood type influences enteric degradation of a cell surface antigen on *Escherichia coli* 086. Gastroenterology 73:37–41
Culling CFA (1974) Handbook of histopathological and histochemical technique. Butterworths, London, pp 232–233
Cuvelier CA, Quatacker J, Mieltans H, De Vos M, Veys E, Roels HJ (1994) M-cells are damaged and increased in number in inflamed human ileal mucosa. Histopathology 24:417–426
Damjanov I (1987) Lectin cytochemistry and histochemistry. Lab Invest 57:5–19
Danguay A, Akif F, Pajak B, Gabius H-J (1994) Contribution of carbohydrate histochemistry to glycobiology. Histol Histopathol 9:155–171
Dayal Y (1991) Neuroendocrine cells of the gastrointestinal tract: introduction and historical perspective. In: Dayal Y (ed) Endocrine pathology of the gut and pancreas. CRC Press, Boca Raton, pp 1–31
Decaens C, Bara J, Rosa B, Daher N, Burtin P (1983) Early oncofoetal antigenic modifications during rat colonic carcinogenesis. Br J Cancer 43:355–362
De Sousa MAB, Parrot DMV, Pantelouris EM (1969) The lymphoid tissues in mice with congenital aplasia of the thymus. Clin Exp Immunol 4:637–644
Dekker J, Aelmans PH, Strous GJ (1991) The oligomeric structure of rat and human gastric mucins. Biochem J 277:423–427
Dorschkind K, Keller GM, Phillips RA, Miller RG, Bosma GC, O'Toole M, Bosma GC (1984) Functional status of cells from lymphoid and myeloid tissues in mice and severe combined immunodeficient disease. J Immunol 132:1804–1808
Dowling RH, Riecken EO, Laws JW, Booth CC (1969) The intestinal response to high bulk feeding in the rat. Clin Sci 32:1–9
Drury RAB, Wallington EA (1980) Carbohydrates and mucosubstances. In: Carleton's histological technique, 5th edn. Oxford University Press, Oxford, pp 232–259
Ducluzeau R (1984) Microbial interactions in the digestive tract. In: Coates ME, Gustafsson BE (eds) The germ-free animal in biomedical research. Laboratory Animals, London, pp 141–154
Ducluzeau R, Raibaud P (1979) Ecologie microbienne du tube digestif. Masson, New York, pp 64–65
Efendic S, Mattsson OE (1978) Effect of somatostatin on intestinal motility. Acta Radiol 19:348–352
Egea G, Franci C, Gambus G, Lesuffleur T, Zweibaum A, Real FX (1993) *cis*-Golgi resident proteins and *O*-glycans are abnormally compartmentalized in the RER of colon cancer cells. J Cell Sci 105:819–830
Ehsanullah M, Filipe MI, Gazzard B (1982) Mucin secretion in inflammatory bowel disease: correlation with disease activity and dysplasia. Gut 23:485–489
Ehsanullah M, Nauton Morgan M, Filipe MI, Gazzard B (1985) Sialomucins in the assessment of dysplasia and cancer-risk patients with ulcerative colitis treated with colectomy and ileorectal anastomosis. Histopathology 9:223–235
Ensinck JW, Vogel RE, Laschansky EC, Francis BH (1990) Effect of ingested carbohydrate, fat and protein on the release of somatostatin-28 in humans. Gastroenterology 98:633–638
Ermak TH, Owen RL (1987) Phenotype and distribution of T lymphocytes in Peyer's patches of athymic mice. Histochemistry 87:321–325
Etzler ME, Branstrator ML (1974) Differential localization of cell surface and secretory components in rat intestinal epithelium by use of lectins. J Cell Biol 62:329–343
Falk P, Roth KA, Gordon GI (1994) Lectins are sensitive tools for defining the differentiation programs of mouse gut epithelial cells lineages. Am J Physiol 266:G987–G1003
Filipe MI (1969) Value of histochemical reactions for mucosubstances in the diagnosis of certain pathological conditions of the colon and rectum. Gut 10:577–586
Filipe MI (1979) Mucins in the human gastrointestinal epithelium: a review. Invest Cell Pathol 2:195–216
Filipe MI, Branfoot AC (1974) Abnormal patterns of mucus secretion in apparently normal mucosa of large intestine with carcinoma. Cancer 34:282–290

Filipe MI, Fenger C (1979) Histochemical characteristics of mucins in the small intestine. A comparative study of normal mucosa, benign epithelial tumours and carcinoma. Histochem J 11:277–287
Filipe MI, Lake BD (1983) Histochemistry in pathology. Churchill Livingstone, Edinburgh
Finegold SM, Sutter VL, Mathisen GE (1983) Microbial composition and development. 1. Normal indigenous intestinal flora. In: Hentges DJ (ed) Human intestinal microflora in health and disease. Academic Press, New York, pp 3–31
Fischer J, Klein PJ, Vierbuchen M, Fischer R, Uhlenbruck G (1983) Lectin binding properties of glycoproteins in cells of normal gastric mucosa and gastric cancers: a comparative histochemical and biochemical study. Cancer Detect Prev 6:137–147
Fischer J, Klein PJ, Vierbuchen M, Skutta B, Uhlenbruck G, Fischer R (1984) Characterization of glycoconjugates of human gastrointestinal mucosa by lectins. I. Histochemical distribution of lectin binding sites in normal alimentary tract as well as benign and malignant gastric neoplasms. J Histochem Cytochem 32:681–689
Florey HW, Wright RD, Jennings MA (1941) Secretions of intestine. Physiol Rev 21:36–69
Florey HW (1955) Mucin and protection in the body. Proc R Soc Biol Lond B 143:147–159
Florey HW (1962) The secretion and function of intestinal mucins. Gastroenterology 43:326–329
Florey HW (1970) General pathology, 4th edn. Lloyd-Luke, London, p 195
Fontaine N, Meslin JC (1994) Spectrocolorimetric assay for the 3 types of gastrointestinal mucins using histochemical stains. Reprod Nutr Dev 34:237–247
Fontaine N, Meslin JC, Lory S, Andrieux C (1996) Intestinal mucin distribution in the germ-free rat and in the heteroxenic rat harbouring a human bacterial flora – effect of inulin in the diet. Br J Nutr 75:881–892
Forstner JF (1978) Intestinal mucins in health and disease. Digestion 17:234–263
Forstner G, Wesley A, Forstner J (1982) Mucus in health and disease. Plenum, New York, pp 199–224
Freeman HJ, Etzler ME, Garrido AB, Kim YS (1978) Alterations in cell surface components of adapting rat small intestinal epithelium: studies with lectins after massive proximal jejunoileal resection and jejunoileal transposition. Gastroenterology 75:1066A
Freeman HJ, Lotan R, Kim YS (1980) Application of lectins for detection of goblet cell glycoconjugate differences in proximal and distal colon of the rat. Lab Invest 42:405–412
Freeman JA (1966) Goblet cell fine structure. Anat Rec 154:121–148
Gaastra W, De Graff FK (1982) Host-specific fimbrial adhesins of non-invasive enterotoxigenic *Escherichia coli* strains. Microbiol Rev 46:229–261
Gebert A (1997) The role of M cells in the production of mucosal membranes. Histochem Cell Biol 108:455–470
Gebert A, Hach G (1993) Differential binding of lectins to M cells and enterocytes in the rabbit caecum. Gastroenterology 105:1350–1361
Gebert A, Hach G, Bertels H (1992) Co-localization of vimentin and cytokeratins in M-cells of rabbit gut-associated lymphoid tissue (GALT). Cell Tissue Res 269:331–340
Gebert A, Schumacher U (1991) Selektive Markierung von M-Zellen durch Fucose- und *N*-acetyl Galactosamin-spezifische Lektine in den caecalen lymphatischen Plaques des Kaninchens. Verh Anat Ges 85:117–118
Giannasca PJ, Giannasca KT, Falk P, Gordon GI, Neutra MR (1994) Regional differences in glycoconjugates of intestinal M cells in mice: potential targets for mucosal vaccines. Am J Physiol 267:G1108–G1121
Giannasca PJ, Neutra MR (1994) Interactions of microorganisms with intestinal M cells: mucosal invasion and induction of secretory immunity. Infect Agent Dis 2:242–248
Goda T, Takase S (1994) Effects of dietary fat content on microvillus in rat jejunum. J Nutr Sci Vitaminol (Tokyo) 40:127–136
Goodlad RA, Lenton W, Ghatei MA, Adrian TE, Bloom SR, Wright NA (1987) Proliferate effects of 'fibre' on the intestinal epithelium: relationship to gastrin, enteroglucagon and PYY. Gut 28:221–226
Goodlad RA, Lee CY, Wright NA (1992) Cell proliferation in the small intestine and colon of intravenously fed rats: effects of urogastrone-epidermal growth factor. Cell Prolif 25:393–404
Goodlad RA, Ratcliffe BR, Fordham JP, Wright NA (1989) Does dietary fibre stimulate intestinal epithelial cell proliferation in germ-free rats? Gut 30:820–825

Goodlad RA, Wright NA (1983) The effects of addition of cellulose or kaolin to an elemental diet on intestinal cell proliferation in the mouse. Br J Nutr 50:91–98
Goodman MJ, Kent PW, Truelove SC (1977) Glucosamine synthetase activity in the colonic mucosa in ulcerative colitis and Crohn's disease. Gut 18:219–228
Gorbach SL (1975) Intestinal microflora in idiopathic inflammatory bowel disease. In: Kirsner B, Shorter RG (eds) Inflammatory bowel disease. Lea and Febiger, Philadelphia, pp 47–61
Gordon JI, Schmidt GH, Roth KA (1992) Studies of intestinal stem cells using normal, chimeric, and transgenic mice. FASEB J 6:3039–3050
Gordon HA, Wostman BS (1973) Chronic mild diarrhoea in germ-free rodents: a model portraying host-flora synergism. In: Heneghan JB (ed) Germ-free research. Academic Press, New York, pp 593–601
Gorelick FS, Sarras MP, Jamieson JD (1982) Regional differences in lectin binding to colonic epithelium by fluorescent and electron microscopy. J Histochem Cytochem 30:1097–1108
Greaves P, Filipe MI, Branfoot AC (1980) Transitional mucosa and survival in human colorectal cancer. Cancer 46:764–770
Greco V, Lauro G, Fabriani A (1967) Histochemistry of the colonic epithelial mucins in normal subjects and in patients with ulcerative colitis. A quantitative and histophotometric investigation. Gut 8:491–496
Griebel PJ, Hein WR (1996) Expanding the role of Peyer's patches in B-cell ontogeny. Immunol Today 30:30–38
Grimelius L (1968) A silver nitrate stain for α_2 cells of human pancreatic islets. Acta Soc Med Upsala 73:243–270
Guitteny AF, Fouque B, Mougin C, Teoule R, Bloch B (1988) Histological detection of messenger RNAs with biotinylated synthetic oligonucleotide probes. J Histochem Cytochem 36:563–571
Gustafsson BE, Maunsbach AB (1971) Ultrastructure of the enlarged caecum in germ-free rats. Z Zellforsch Mikrosk Anat 120:555–578
Habib NA, Dawson PM, Krausz T, Blount MA, Kersten D, Wood CB (1986) A study of histochemical changes in mucus from patients with ulcerative colitis, Crohn's disease and diverticular disease of the colon. Dis Colon Rectum 29:15–17
Harbers LH, George JR, Reeves RD (1980) Digestion of lettuce in rats observed by scanning electron-microscopy. Nutr Rep Int 21:681–688
Heneghan JB (1984) Physiology of the alimentary tract. In: Coates ME, Gustafsson BE (eds) The germ-free animal in biomedical research. Laboratory Animals, London, pp 169–191
Heidenhain R (1870) Untersuchungen über den Bau der Labdrüsen. Arch Mikrosk Anat 6:368–406
Heidsiek JG, Hyde DM, Plopper CG, St. George JA (1987) Quantitative histochemistry of mucosubstance in tracheal epithelium of the Macaque monkey. J Histochem Cytochem 35:435–442
Hein WR, Dudler L, Mackay CR (1989) Surface expression of differentiation antigens on lymphocytes in the ileal and jejunal Peyer's patches of lambs. Immunology 68:365–370
Heneghan JB (1979) Enterocyte kinetics, mucosal surface area and mucus in gnotobiotes. In: Fliedner T, Heit H, Niethammer D, Pflieger H (eds) Clinical and experimental gnotobiotics. Fischer, Stuugart, pp 19–27
Hilbert DM, Anderson AO, Holmes KL, Rudikoff S (1994) Long-term reconstitution of scid mice suggests self-renewing B and T cell populations in peripheral and mucosal tissues. Transplantation 58:466–475
Hill RRH, Cowley HM, Andremont A (1990) Influence of colonizing micro-flora on the mucin histochemistry of the neonatal mouse colon. Histochem J 22:102–105
Ho SB, Niehans GA, Lyftogt C, Yan PS, Cherwitz DL, Gum ET, Dahiya R, Kim YS (1993) Heterogeneity of mucin gene expression in normal and neoplastic tissues. Cancer Res 53:641–651
Höfer D, Drenckhahn D (1996) Cytoskeletal markers allowing discrimination between brush cells and other epithelial cells of the gut including enteroendocrine cells. Histochem Cell Biol 98:405–412
Hoskins LC (1978) Degradation of mucus glycoproteins in the gastrointestinal tract. In: Horowitz M, Pigman W (eds) The glycoconjugates. Academic Press, New York, pp 235–253
Hoskins LC (1984) Mucin degradation by enteric bacteria: ecological aspects and implications for bacterial attachment to gut mucosa. In: Boedeker EC (ed) Attachement of organisms to the gut mucosa. CRC Press, Boca Raton, pp 51–67

Hoskins LC, Boulding ET (1981) Mucin degradation in human colon ecosystems. Evidence for the existence and role of the bacterial subpopulations producing glycosidases as extracellular enzymes. J Clin Invest 67:163–172
Hoskins LC, Agustines M, McKee WB, Boulding ET, Kriaris M, Niedermeyer G (1985) Mucin degradation in human colon ecosystems. Isolation and properties of faecal strains that degrade ABH blood group antigens and oligosaccharides from mucin glycoproteins. J Clin Invest 75:944–953
Howell WH (1926) A textbook of physiology for medical students and physicians. Saunders, Philadelphia, p 802
Hsu SM, Raine L, Fanger H (1981) Use of avidin-biotin-peroxidase complex (ABC) in immunoperoxidase techniques: a comparison between ABC and unlabelled antibody (PAP) procedures. J Histochem Cytochem 29:577–580
Husband AJ, Gowans JL (1978) The origin and antigen-dependent distribution of IgA-containing cells in the intestine. J Exp Med 148:1146–1160
Husebye E, Hellström PM, Midvedt T (1990) The role of the normal microbial flora in the control of small intestinal motility. Microecol Ther 20:384–394
Ikeda RM, Gershwin ME (1978) In: Gershwin ME, Cooper EL (eds) Animal models of comparative and developmental aspects of immunity and diseases. Pergamon, pp 201–210
Inman LR, Cantey JR (1984) Specific adherence of *Escherichia coli* (strain RDEC-1) to membranous (M) cells of the Peyer's patch in *Escherichia coli* diarrhoea in the rabbit. J Clin Invest 71:1–8
Ishikawa K, Satoh Y, Tanaka H, Ono K (1986) Influence of conventionalization on small-intestinal mucosa of germ-free wistar rats: quantitative light microscopic observations. Acta Anat 127:296–302
Itagaki S, Perfumo CJ, Petruccelli MA, Doi K (1994) Lectin histochemical changes of goblet cell mucin in rabbit mucoid enteropathy. Lab Anim Sci 44:82–84
Jaan K, Hoschützky H (1991) Nature and organisation of adhesins. Curr Top Microbiol Immunol 156:55–70
Jacobs LR (1983) Effects of dietary fibre on mucosal growth and cell proliferation in the small intestine of the rat: a comparison of oat bran, pectin, and guar with total fibre deprivation. Am J Clin Nutr 37:954–960
Jacobs LR (1985) Differential effects of dietary fibres on rat intestinal circular muscle cell size. Dig Dis Sci 30:247–252
Jacobs LR, De Fontes D, Cox KL (1983) Cytochemical localization of small intestinal glycoconjugates by lectin histochemistry in controls and subjects with cystic fibrosis. Dig Dis Sci 28:422–428
Jacobs LR, Huber WP (1985) Regional distribution and alterations of lectin binding to colorectal mucins in mucosal biopsies from controls and subjects with inflammatory bowel disease. J Clin Invest 75:112–118
Jacobs LR, Lupton JR (1984) Effects of dietary fibre on rat bowel mucosal growth and cell proliferation. Am J Physiol 246:G378–G385
Jacobs LR, Schneeman BO (1981) Effects of dietary wheat bran on rat colonic structure and mucosal cell growth. J Nutr 111:798–803
Jenkins AP, Thomson RPH (1989) Trophic effect of Efamol on the rat small intestinal mucosa. Clin Sci 77:555–559
Jepson MA, Mason CM, Bennett MK, Simmons NL, Hirst BH (1992) Co-expression of vimentin and cytokeratins in M cells of rabbit intestinal lymphoid follicle-associated epithelium. Histochem J 24:33–39
Jepson MA, Clark MA, Simmons NL, Hirst BH (1993) Epithelial M cells in the rabbit caecal lymphoid patch display distinctive surface characteristics. Histochemistry 100:441–447
Jepson MA, Mason CM, Simmons NL, Hirst BH (1995) Enterocytes in the follicle-associated epithelia of rabbit small intestine display distinctive lectin-binding properties. Histochemistry 103:131–134
Johnson IT, Gee JM, Brown JC (1988) Plasma enteroglucagon and small bowel cytokinetics in rats fed soluble nonstarch polysaccharides. Am J Clin Nutr 47:1004–1009
Jones GW, Abrams GD, Freter R (1976) Adhesive properties of *Vibrio cholerae*: adhesion to isolated rabbit brush border membranes and haemagglutinating activity. Infect Immun 14:232–239
Kemper AC, Specian RD (1991) Rat small intestinal mucins: a quantitative analysis. Anat Rec 229:219–226

Kim YS, Issacs R (1979) Glycoprotein metabolism in inflammatory and neoplastic diseases of the human colon. Cancer Res 35:2092–2097

Kirchgessner CU, Patil CK, Evans JW, Cuomo CA, Fried LM, Carter T, Oettinger MA, Brown JM (1995) DNA-dependent kinase (P350) as a candidate gene for the murine scid defect. Science 267:1178–1183

Kirkwood BR (1992) Essentials of medical statistics. Blackwell Scientific Publications, Oxford, pp 41–45

Kraehenbuhl J-P, Neutra MR (1992) Molecular and cellular basis of immune protection of mucosal surfaces. Physiol Rev 72:853–879

Krause WJ, Yamada J, Cutts JH (1985) Quantitative distribution of enteroendocrine cells in the gastrointestinal tract of the adult opossum, *Didelphis virginiana*. J Anat 140:591–605

Kritchevsky D (1988) Dietary fibre. Annu Rev Nutr 8:301–328

Kroese FGM, Timens W, Nieuwenhuis P (1990) Germinal centre reaction and B lymphocytes: morphology and function. In: Grundmann E, Vollmer E (eds) Reaction patterns of the lymph node, part 1. Cell types and functions. Verlag, Berlin, pp 103–148

Lamont JT (1992) Mucus: the front line of intestinal mucosal defense. Ann NY Acad Sci 664:190–201

Lamont JT, Ventola AS, Maull EA, Szabo S (1983) Cystamine and prostaglandin F stimulate rat gastrin mucin release. Gastroenterology 84:306–313

Lee A, O'Rourke JL, Barrington PJ, Trust TJ (1986) Mucus colonization as a determinant of pathogenicity in intestinal infection of *Campylobacter jejuni*: a mouse caecal model. Infect Immun 51:536–546

Lee DY, Chapkin RS, Lupton JR (1993) Dietary fat and fibre modulate colonic cell proliferation in an interactive site-specific manner. Nutr Cancer 20:107–118

Lefevre ME, Olivo R, Vanderhoff JW, Joel DD (1978) Accumulation of latex in Peyer's patches and its subsequent appearance in villi and mesenteric lymph nodes. Proc Soc Exp Biol Med 159:298

Lev R, Lance P, Camara P (1985) Histochemical and morphological studies of mucosa bordering rectosigmoid carcinomas: comparisons with normal, diseased and malignant colonic epithelium. Hum Pathol 16:151–161

Lev R, Spicer SS (1964) Specific staining of sulphate groups with alcian blue at low pH. J Histochem Cytochem 12:309

Lhuillery C, Demarne Y, Dubos F, Galpin JV, Ducluzeau R, Raibaud P (1981) Fatty acid composition of lipids in the maternal diet and establishment of a Lactobacillus sp. strain in the digestive tract of suckling gnotobiotic mice and rats. Am J Clin Nutr 34:1513–1519

Lindstedt G, Lindstedt S, Gustafsson BE (1965) Mucus in intestinal contents of germ-free rats. J Exp Med 121:201–213

Listinsky CM, Ridell RH (1981) Patterns of mucin secretion in neoplastic and non-neoplastic diseases of the colon. Hum Pathol 12:923–929

Loesche WJ (1968) Protein and carbohydrate of caecal contents of gnotobiotic rats and mice. Proc Soc Exp Biol Med 128:195–199

Lundin E, Zhang J-X, Huang C-B, Reuterving C-O, Hallmans G, Nygren C, Stenling R (1993) Oat bran, rye bran, and soybean hull increase goblet cell volume density in the small intestine of the golden hamster. A histochemical and stereological light-microscopic study. Scand J Gastroenterol 28:15–22

MacDermott RP, Donaldson RM, Trier JS (1974) Glycoprotein synthesis and secretion by mucosal biopsies of rabbit colon and human rectum. J Clin Invest 54:545–554

Mallett AK, Bearne CA, Rowland IR, Farthing MJG, Cole CB, Fuller R (1987) The use of rats associated with a human faecal flora as a model for studying the effects of diet on the human gut microflora. J Appl Bacteriol 63:39–45

Mantle M, Allen A (1989) Gastrointestinal mucus. In: Davison JS (ed) Gastrointestinal secretion. Wright, London, pp 202–223

Mantle M, Husar SD (1994) Binding of *Yersinia enterocolitica* to purified, native small intestinal mucins from rabbits and human involves interactions with the mucin carbohydrate moiety. Infect Immun 62:1219–1227

Mantle ML, Basaraba BC, Peacock C, Gall DH (1989) Binding of *Yersinia enterocolitica* to rabbit intestinal brush border membranes, mucus and mucin. Infect Immun 57:3239–3299

Masson P (1914) Le Glande endocrine de l'intestin chez l'homme. Crit Rev Acad Sci (Paris) 158:59–61

Mathers JC, Kennard J (1993) Gastrointestinal responses to oats consumption in young adult and elderly rats: digestion, large bowel fermentation and crypt cell proliferation rates. Br J Nutr 70:567–584
Maxton DG, Cynk EU, Jenkins AP, Thompson RPH (1989) Effects of dietary fat on the small intestinal mucosa. Gut 30:1252–1255
Menguy R, Thompson AE (1967) Regulation of secretion of mucus from the gastric antrum. Ann NY Acad Sci 140:797–803
Merzel J, Almeida ML (1973) Maturation of goblet cells of the mouse small intestine as shown by the uptake of 35S-sodium sulfate. Anat Rec 177:519–524
Merzel J, Leblond CP (1969) Origin and renewal of goblet cells in the epithelium of the mouse small intestine. Am J Anat 124:281–305
Meslin JC, Andrieux C, Sakata T, Beaumatin P, Bensaada M, Popot F, Szylit O, Durand M (1993) Effects of galacto-oligosaccharide and bacterial status on mucin distribution in mucosa and on large intestine fermentation in rats. Br J Nutr 69:903–912
Midtvedt T, Carlstedt-Duke B, Höverstad T, Midtvedt A-C, Norin KE, Saxerholt H (1987) Establishment of a biochemically active intestinal ecosystem in ex-germfree rats. Appl Environ Microbiol 53:2866–2871
Moré J, Fioramonti F, Benazet F, Bueno L (1987) Histochemical characterization of glycoproteins present in jejunal and colonic goblet cells of pigs on different diets. A biopsy study using chemical methods and peroxidase-labelled lectins. Histochemistry 87:189–194
Moore WEC, Cato EP, Holdeman LV (1978) Some current aspects in intestinal bacteriology. Am J Clin Nutr 31:S33–S42
Morrissey SM, Tymvios MC (1987) Acid mucins in human intestinal goblet cells. J Pathol 126:197–208
Mowry RW (1963) The special value of methods that colour both acidic and vicinyl hydroxyl groups in the histochemical study of mucins, with revised directions for the colloidal iron stain, the use of Alcian Blue 8Gx and their combination with the periodic acid Schiff reaction. Ann NY Acad Sci 106:402–423
Nagura H, Ohtani H, Masuda T, Kimura M, Nakamura S (1991) HLA-DR expression on M cells overlying Peyer's patches is a common feature of human small intestine. Acta Pathol (Jpn) 41:818–823
Neutra MR, Forstner JF (1987) Gastrointestinal mucus, synthesis, secretion, and function. In: Johnson LR (ed) Physiology of the gastrointestinal tract, 2nd edn. Raven, New York, pp 975–1009
Neutra MR, Frey A, Kraehenbuhl J-P (1996b) Epithelial M cells: gateways for mucosal infection and immunization. Cell 86:345–348
Neutra MR, Kraehenbuhl J-P (1992) Transepithelial transport and mucosal defence. I. The role of M cells. Trends Cell Biol 2:134–138
Neutra MR, O'Malley LJ, Specian RD (1982) Regulation of intestinal goblet cell secretion. II. A survey of potential secretagogues. Am J Physiol 242:G380–G387
Neutra MR, Phillips TL, Mayer EL, Fishkind DJ (1987) Transport of membrane-bound macromolecules by M cells in follicle-associated epithelium of rabbit Peyer's patch. Cell Tissue Res 247:537–546
Neutra MR, Pringault E, Kraehenbuhl J-P (1996) Antigen sampling across epithelial barriers and induction of mucosal immune responses. Annu Rev Immunol 14:275–300
Nieuwenhuis P (1971) On the origin and fate of immunologically competent cells. Wolter-Nordhoff, Groningen
Norin KE, Gustafsson BE, Lindblad BS, Midvedt T (1985) The establishment of some microflora-associated biochemical characteristics in faeces from children during the first years of life. Acta Paediatr Scand 74:207–212
Ofek I, Sharon N (1988) Lectinophagocytosis: a molecular mechanism of recognition between cell surface sugars and lectins in the phagocytosis of bacteria. Infect Immun 56:539–547
Ohara S, Byrd JC, Gum JR, Kim YS (1994) Biosynthesis of two distinct types of mucin in HM3 colon cancer cells. Biochem J 297:509–516
O'Leary AD, Sweeny EC (1986) Lymphoglandular complexes of the colon: structure and distribution. Histopathology 10:267–283

Osmond DG, Nossal GJV (1974) Differentiation of lymphocytes in mouse bone marrow. II. Kinetics of maturation and renewal of antiglobulin-binding cells studied by double labelling. Cell Immunol 13:132–145
Owen RL (1977) Sequential uptake of horseradish peroxidase by lymphoid follicle epithelium of Peyer's patches in the normal unobstructed mouse intestine: an ultrastructural study. Gastroenterology 72:440–451
Owen RL (1994) M cells – entryways of opportunity for enteropathogens. J Exp Med 180:7–9
Owen RL, Bhalla DK (1983) Cytochemical analysis of alkaline phosphatase and esterase activities and of lectin-binding and anionic sites in rat and mouse Peyer's patch M cells. Am J Anat 168:199–212
Owen RL, Ermak TH (1990) Structural specialisations for antigen uptake and processing in the digestive tract. Springer Semin Immunopathol 12:139–152
Owen RL, Jones AL (1974) Epithelial specialization within human Peyer's patches: an ultrastructural study of intestinal lymphoid follicles. Gastroenterology 66:189–203
Park CM, Reid PE, Owen DA, Volz D, Dunn WL (1987) Histochemical studies of epithelial cell glycoproteins in normal rat colon. Histochem J 19:546–554
Paulus U, Loeffler M, Zeidler J, Owen G, Potten S (1993) The differentiation and lineage development of goblet cells in the murine small intestinal crypt: experimental and modelling studies. J Cell Sci 106:473–484
Pearse AGE (1985) Histochemistry: theoretical and applied, vol 2, 4th edn. Churchill Livingstone, Edinburgh
Pearse AGE, Polak JM (1978) The diffuse neuroendocrine system and the APUD concept. In: Bloom SR (ed) Gut hormones. Churchill-Livingstone, Edinburgh, pp 33–39
Pell JD, Brown JC, Johnson IT (1994) Polyunsaturated fatty-acids of the n-3 series influence intestinal crypt cell production in rats. Carcinogenesis 15:1115–1119
Pell JD, Johnson IT, Goodlad RA (1995) The effects of and interactions between fermentable dietary fibre and lipid in germfree and conventional mice. Gastroenterology 108:1745–1752
Peumans WJ, Van Damme EJM (1994) Recent advances in the purification and characterization of plant lectins and their introduction as tools. In: Van Driessche E, Fischer J, Beeckmans S, Bøg-Hansen TC (eds) Lectins: biology, biochemistry, clinical biochemistry, vol 10. Textop, Hellerup, pp 118–127
Phillips TE, Phillips TH, Neutra MR (1984) Regulation of intestinal goblet cell secretion. II. A survey of potential secretagogues. Am J Physiol 247:G674–G681
Phillips TE, Wilson J (1993) Morphometric analysis of mucous granule depletion and replenishment in rat colon. Dig Dis Sci 38:2299–2304
Pigman W, Moschera J, Weis M, Tettamanti G (1973) The occurrence of repetitive glycopeptide sequences in bovine submaxillary glycoprotein. Eur J Biochem 32:148–154
Piotrowski J, Murty VL, Czajkowski A, Slomiany A, Yotsumoto F, Majka J, Slomiany BL (1994) Association of salivary bacterial aggregating activity with sulfomucin. Biochem Mol Biol Int 32:713–721
Plendl J, Schönleber B, Schmahl W, Schumacher U (1989) Comparison of the unmasking of lectin receptors by neuraminidase and by enzyme free buffer alone. J Histochem Cytochem 37:1743–1744
Podolsky DK, Fournier DA, Lynch KE (1986) Development of anti-human colonic mucin monoclonal antibodies. Characterization of multiple mucin species. J Clin Invest 77:1251–1262
Podolsky DK, Isselbacher KJ (1983) Composition of human colonic mucin. Selective alteration in inflammatory bowel disease. J Clin Invest 72:142–153
Pollard M, Sharon N (1970) Responses of the Peyer's patches in germ-free mice to antigenic stimulation. Infect Immun 2:96–100
Polak JM (1989) Endocrine cells of the gut. In: Schultz G (ed) Handbook of physiology, sect 6. The gastrointestinal system, vol 2. Neural and endocrine biology. American Physiological Society, Bethesda, pp 79–96
Posel P, Schumacher U, Frick H, Welsch U (1988) Histochemistry of mucosubstances in Brunner's gland cells and duodenal goblet cells of the two new World monkey species (*Saimiri sciureus* and *Saguinus fuscicollis*). Histochemistry 88:327–332
Proust JE, Tchaliovska SD, Ter Minassian Sar L (1984) Mucin thin film as a model of the tear film rupture. Science 98:319

Radwan KA, Oliver MG, Specian RD (1990) Cytoarchitectural reorganization of rabbit colonic goblet cells during baseline secretion. Anat Rec 189:365–376
Rautenberg K, Cichon C, Heyer G, Demel M, Schmidt MA (1996) Immunocytochemical characterization of the follicle-associated epithelium of Peyer's patches: anti-cytokeratin 8 antibody (clone 4.1.18) as a molecular marker for rat M cells. Eur J Cell Biol 71:363–370
Ravdin JI, John JE, Johnston LI, Innes D, Guerrant RL (1985) Adherence of *Entamoeba histolytica* trophozoites to rat and human colonic mucosa. Infect Immun 48:292–297
Reichman FM, Bosma MJ, Hardy RR (1993) B-lineage cells in mu-transgenic scid mice proliferate in response to IL-7 but fail to show evidence of immunoglobulin light chain gene arrangement. Int Immunol 5:303–310
Reid PE, Culling CFA, Dunn WL (1974) The histochemical interpretation of the complex results of methylation upon gastrointestinal tract mucins with special reference to the periodic acid-Schiff reactivity. J Histochem Cytochem 22:986–991
Reid PE, Culling CFA, Dunn WL, Ramey CW, Clay MG (1984) Chemical and histochemical studies of normal and diseased human gastrointestinal tract. I. A comparison between histologically normal colon, colonic tumours, ulcerative colitis and diverticular disease of the colon. Histochem J 16:235–251
Reid PE, Walker DC, Terpin T, Owen DA (1988) Histochemical studies of the colonic epithelial glycoproteins of the normal rabbit. Histochem J 20:535–550
Reynaud C-A, Mackay CR, Müller RG, Weill J-C (1991) Somatic generation of diversity in a mammalian lymphoid organ: the sheep ileal Peyer's patches. Cell 64:995–1005
Reynolds JD, Morris B (1983) The evolution and involution of Peyer's patches in fetal and neonatal sheep. Eur J Immunol 13:627–635
Rhodes JM (1989) Colonic mucus and mucosal glycoproteins: the key to colitis and cancer? Gut 26:761–765
Rose MC (1989) Characterization of human tracheobronchial mucin glycoproteins. Methods Enzymol 179:3–17
Roth KA, Hertz JM, Gordon JI (1990). Mapping enteroendocrine cell populations in transgenic mice reveals an unexpected degree of complexity in cellular differentiation within the gastrointestinal tract. J Cell Biol 110:1791–1801
Roth KA, Kim S, Gordon JI (1992) Immunological studies suggest two pathways for enteroendocrine cell differentiation in the colon. Am J Physiol 263:G174–G180
Roussel P, Lamblin G, Houdret N, Lhermitte M, Slayter HS (1984) Conformation of human mucus glycoproteins observed by electron microscopy. Biochem Soc Trans 12:617–618
Roy MJ (1987) Precocious development of lectin (*Ulex europaeus* agglutinin I) receptors in the dome epithelium of gut-associated lymphoid tissues. Cell Tissue Res 248:483–489
Roy MJ, Ruiz AR (1986) Dome-associated epithelial M cells dissociated from rabbit gut-associated lymphoid tissues. Am J Vet Res 47:2577–2583
Roy MJ, Ruiz A, Varvayanis M (1987) A novel antigen is common to the dome epithelium of gut- and bronchus associated lymphoid tissues. Cell Tissue Res 248:635–644
Roy MJ, Varvayanis M (1987) Development of dome epithelium in gut-associated lymphoid tissues: association of IgA with M cells. Cell Tissue Res 248:645–651
Rozee KR, Cooper D, Lam K, Costerton JW (1982) Microbial flora of the mouse ileum mucous layer and epithelial surface. Appl Environ Microbiol 43:1451–1463
Rubio CA, Saraga EP, Lindholm J (1990) Improved method for mapping gastric intestinal metaplasia using selective histochemical morphometry. Anal Quant Cytol 12:122–127
Rumney CJ, Rowland IR (1992) In vivo and in vitro models of the human colonic flora. Crit Rev Food Sci Nutr 31:299–331
Sacerdotti (1894) Über die Entwicklung der Schleimzellen des Magendarmkanales. Int Mschr Anat Physiol 11:501–514
Sakata T (1987) Stimulatory effect of short-chain fatty acids on epithelial cell proliferation in the rat intestine: a possible explanation for trophic effects of fermentable fibre, gut microbes and luminal trophic factors. Br J Nutr 58:95–103
Sakata T, Von Engelhardt W (1981) Influence of short-chain fatty acids and osmolarity on mucin release in the rat colon. Cell Tissue Res 219:371–377

Sakata T, Yajima T (1984) Influence of short chain fatty acids on the epithelial cell division of digestive tracts. Q J Exp Physiol 69:639–648
Sata T, Roth J, Zuber C, Stamm B, Heitz PU (1991) Expression of α2,6 linked sialic acid residues in neoplastic but not in normal human colonic mucosa. A lectin-gold cytochemical study with *Sambucus nigra* and *Maackia amurensis* lectins. Am J Pathol 139:1435–1448
Savage DC (1977) Microbial ecology of the gastrointestinal tract. Annu Rev Microbiol 31:107–133
Savage DC (1978) Factors involved in colonization of the gut epithelial surface. Am J Clin Nutr 31 [Suppl 10]:S131–S135
Savage DC (1980) Adherence of normal flora to mucosal surfaces. In: Beachey EH (ed) Bacterial adherence. Chapman and Hall, London, pp 31–60
Savage DC (1983) Morphological diversity among members of the gastrointestinal microflora. Int Rev Cytol 82:305–334
Savage DC (1984) Present view of the normal flora. In: Coates ME, Gustafsson BE (eds) The germ-free animal in biomedical research. Laboratory Animals, London, pp 141–154
Savage DC (1989) The normal human microflora-composition. In: Grubb R, Midtvedt T, Norin E (eds) The regulatory and protective role of the normal microflora. Macmillan, Basingstoke, pp 3–18
Savidge TC (1996) The life and times of an intestinal M cell. Trends Microbiol 4:301–306
Schauer R (1985) Sialic acids and their role as biological masks. Trends Biochem Sci 10:357–360
Schneeman BO, Richter BD, Jacobs LR (1982) Response to dietary wheat bran in the exocrine pancreas and intestine of rats. J Nutr 112:283–286
Schumacher U, Adam E (1994) A dot blot technique of standardisation of staining in glycosaminoglycan histochemistry: Alcian Blue, its analogues, and diamine methods. Biotech Histochem 69:18–24
Schumacher U, Madry H, Adam E, Peumans WJ, Damme EJM van, Grant G, Bardocz S, Pusztai A (1993) Analysis of lectin binding sites in the gut of hooded lister rats with special emphasis on recently detected lectins. Acta Histochem 94:163–166
Scott JE, Dorling J (1965) Differential staining of acid glycosaminoglycans (mucopolysaccharides) by Alcian Blue in salt solutions. Histochemie 5:221–223
Shamsuddin AKM, Trump BF (1981) Colon epithelium I. Light microscopic, histochemical and ultrastructural features of normal colon epithelium of male Fischer 344 rats. J Natl Cancer Inst 66:375–388
Sharma R, Schumacher U (1992) Histochemical characterization of carbohydrate residues during the morphogenesis of gastrointestinal and respiratory systems of *Caretta caretta*. Acta Histochem 93:411–432
Sharon N (1987) Bacterial lectins, cell-cell recognition and infectious disease. FEBS Lett 217:145–157
Sharon N, Lis H (1972) Lectins: cell-agglutinating and sugar specific proteins. Science 177:949–958
Sheahan DG, Jervis HR (1976) Comparative histochemistry of gastrointestinal mucosubstances. Am J Anat 146:103–132
Sigleo S, Jackson MJ, Vahouny GV (1984) Effects of dietary fibre constituents on intestinal morphology and nutrient transport. Am J Physiol 246:G34–G39
Silberberg A, Meyer FA (1982) Structure and function of mucus. Mucus in health and disease II. Adv Exp Med Biol 144:53–74
Sircar B, Johnson LR, Lichtenberger LM (1983) Effects of synthetic diets on gastrointestinal mucosal DNA synthesis in rats. Am J Physiol 244:G327–G335
Slayter HS, Lamblin G, Le Treut A, Galabert C, Houdret N, Degand P, Roussel P (1984) Complex structure of human bronchial mucus glycoprotein. Eur J Biochem 142:209–218
Smith AC, Podolsky DK (1987) Biosynthesis and secretion of human colonic mucin glycoproteins. J Clin Invest 80:300–307
Snary D, Allen A, Pain RH (1970) Structural studies on gastric mucoproteins: lowering of molecular weight after reduction with 2-mercaptoethanol. Biochem Biophys Res Commun 40:844–851
Sjölund K, Alumets J, Berg N-O, Håkanson R, Sundler F (1982) Enteropathy of coeliac disease in adults: increased number of enterochromaffin cells in the duodenal mucosa. Gut 23:42–48
Sjölund KG, Sanden R, Håkanson R, Sundler F (1983) Endocrine cells in human intestine: an immunocytochemical study. Gastroenterology 85:1120–1130
Solcia E, Capella C, Buffa R, Usellini L, Fiocca R, Sessa F (1987) Endocrine cells of the digestive system. In: Johnson LR (ed) Physiology of the gastrointestinal tract, 2nd edn. Raven, New York, pp 111–130

Solcia E, Creutzfeldt W, Falkmer S, Fujita T, Greider MH, Grossman MI, Grube D, Håkanson R, Larsson L-I, Lechago J, Lewin K, Polak JM, Rubin W (1981) Human gastroenteropancreatic endocrine-paracrine cells: Santa Monica 1980 classification. In: Grossman MI, Brazier MAB, Lechago J (eds) Cellular basis of chemical messengers in the digestive system. Academic Press, New York, pp 159–165

Specian RD, Neutra MR (1980) Mechanism of rapid mucus secretion in goblet cells is stimulated by acetylcholine. J Cell Biol 85:626–640

Specian RD, Neutra MR (1982) Regulation of intestinal goblet cell secretion. I. Role of parasympathetic stimulation. Am J Physiol 242:G370–G379

Specian RD, Neutra MR (1984) Cytoskeleton of intestinal goblet cells in rabbit and monkey. The theca. Gastroenterology 87:1313–1325

Specian RD, Oliver MG (1991) Functional biology of intestinal goblet cells. Am J Physiol 260:C183–C193

Specian RD, Zhang SJ, Sibley DA, Kemper AC, Oliver MG (1990) Recovery of goblet cells from an accelerated secretory event. Anat Rec 226:97A

Spicer SS (1960) A correlative study of the histochemical properties of rodent acid mucopolysaccharides. J Histochem Cytochem 8:18–33

Spicer SS (1965) Diamine methods for differentiating mucosubstances histochemically. J Histochem 13:211–234

Spicer SS, Schulte BA (1992) Diversity of cell glycoconjugates shown histochemically: a perspective. J Histochem Cytochem 40:1–38

Staffos RO, Rhatigan RM (1977) Benign (nonpolyploid) mucosal changes adjacent to carcinomas of the colon. A light microscopic study of 20 cases. Hum Pathol 8:441–449

Stanley RA, Ram SP, Wilkinson RK, Robertson AM (1986) Degradation of pig gastric and colonic mucins by bacteria isolated from the pig colon. Appl Environ Microbiol 51:1104–1107

Strous GJ, Dekker J (1992) Mucin-type glycoproteins. Crit Rev Biochem Mol Biol 27:57–92

Strunz U, Domschke W, Domschke S (1976) Gastrointestinal motor response to natural motilin and synthetic position 13 substituted motilin analogues. A comparative in vitro study. Scand J Gastroenterol 11:199–203

Sundler F, Ekblad E, Håkanson R (1993) Localization and colocalization of gastrointestinal peptides. In: Brown DR (ed) Gastrointestinal peptides. Springer, Berlin Heidelberg New York, pp 1–27

Tannock GW, Savage DC (1974) Influence of dietary and environmental stress on microbial populations in the murine gastrointestinal tract. Infect Immun 9:591–598

Tasman-Jones C, Owen LR, Jones AL (1982) Semipurified dietary fibre and small bowel morphology in rats. Dig Dis Sci 27:519–524

Teague RH, Fraser D, Clamp JR (1973) Changes in monosaccharide content of mucus glycoproteins in ulcerative colitis. Br Med J 2:645–646

Thomopoulos GN, Schulte BA, Spicer SS (1983) Light and electron cytochemistry of glycoconjugates in the rectosigmoid colonic epithelium of the mouse and rat. Am J Anat 168:239–256

Thornton DJ, Davies JR, Kraayenbrink M, Richardson PS, Sheehan JK, Carlsedt T (1990) Mucus glycoproteins from normal human tracheobronchial secretion. Biochem J 265:179–186

Tsuyama S, Suzuki S, Murata F (1983) The histochemical differences of intestinal gland epithelia in the rat colon with special reference to their glycoconjugates. Acta Histochem Cytochem 16:456–471

Uribe A, Alam M, Johansson O, Midtvedt T, Theodorsson E (1994) Microflora modulates endocrine cells in the gastrointestinal mucosa of the rat. Gastroenterology 107:1259–1269

Vahouny GW, Le T, Satchithanandam S, Cassidy MM (1985) Stimulation of intestinal cytokinetics and mucin turnover in rats fed wheat bran or cellulose. Am J Clin Nutr 41:895–900

Varcellotti JR, Salyers AA, Bullard WS, Wilkins TD (1977) Breakdown of mucin and plant polysaccharides in the human colon. Can J Biochem 55:1190–1196

Warner NL, Szenberg A, Burnet FM (1962) The immunological role of different lymphoid organs in the chicken. Aust J Exp Biol Med Sci 40:373–387

Weibel EW (1979) Stereological methods, vol 1. Academic Press, New York

Wold JK, Midtvedt T, Jeanloz RW (1974) Intestinal glycoproteins in germ-free rats. III. Characterization of a water-soluble glycoprotein fraction. Acta Chem Scand B 28:277–284

Wolf JL, Rubin DH, Finberg R, Kauffman RS, Sharpe AH, Trier JS, Fields BN (1981) Intestinal M-cells: a pathway for entry of reovirus into the host. Science 212:471–472

Zotter S, Hageman PC, Lossnitzer A, Mooi WJ, Higers J (1988) Tissue and tumour distribution of human polymorphic epithelial mucin. Cancer Rev 11:55–101

Subject Index